Jon No CPC/1222/03
17/1/004

AF646262

Himalayan Snow and Glaciers

Concept's Discovering Himalayas Series

- People's Participation in Himalayan Eco-System Development: A Plan for Action/ **M.L. Dewan**
- Nut Fruits for the Himalayas (CDHS)/ **M.L. Dewan, M.C. Nautiyal** and **V.K. Sah**
- Vedic Philosophy and Himalayan Eco-System Development/ **M.L. Dewan** and **B.D. Joshi (Eds)**
- People's Movement for Himalayan Rejuvenation : 2nd Revised and enlarged edition/ **M.L. Dewan**
- Water User's Associations In Andhra Pradesh: A Quick Assessment from the field / **Jasveen Jairath**
- Images of Women in the Folksongs of Garhwal Himalayas / **Anjali Capila**

About the Author

Jagdish Bahadur earned his Master's degree in Physics in 1958 and Ph.D. in 1969. In 1969-70, he received a post-doctoral fellowship from Norwegian Agency for International Development on Glaciology. During this period, he, in collaboration with Danish and Swedish scientists, successfully completed a project on evaluation of annual accumulation on a high and cold glacier.
After his return to India, Dr. Bahadur started taking interest in the organisation of glacier studies and has published several papers on the role of Himalayan glaciers and their contribution to the water resources and productivity aspects of Indian landmass. He has presented many papers at international conferences, organised by UNESCO, IGBP and ICIMOD. Also, he has two books to his credit: *Snow & Glaciers* and Their Contribution to *India's Water Resources* (1992), and *Indian Himalayas: An Integrated View* (2003).

Concept's Discovering Himalayas Series No. 7

HIMALAYAN SNOW AND GLACIERS

Associated Environmental Problems, Progress and Prospects

JAGDISH BAHADUR
Principal Investigator, DST Project on Himalayas,
Vigyan Prasar, New Delhi-110016

CONCEPT PUBLISHING COMPANY, NEW DELHI-110059

All rights reserved. No part of this work may be reproduced, stored in a retrieval system, or transmitted in any form or by any means electronic, mechanical, photocopying, micro-filming recording or otherwise, without the prior written permission of the copyright owner and the publisher.

ISBN 81-8069-091-1

First Published 2004

© Jagdish Bahadur (*b.1938*)

Published and Printed by
Ashok Kumar Mittal
Concept Publishing Company
A/15-16, Commercial Block, Mohan Garden
New Delhi-110059 (INDIA)

Phones : 25351460, 25351794
Fax : 091-11-25357103
Email : publishing@conceptpub.com

Dedicated to the
fond memory of my beloved parents
who always kindly supported me and my activities

दिवोमूलभवततं पृथिव्या अध्युत्ततम्।
तेन सहस्रकाण्डेन परि णः पाहि विश्वतः॥

As rays come down from the Sun, spread themselves on the Earth, and shed lustre, so is the knowledge of God, in the shape of the Vedas, revealed by Him, and spreads all over the Earth. O God, with this Vedic knowledge of innumerable branches, fully protect us from every side.

Atharva Veda

Foreword

Himalayan mountain ranges influence global air circulations. The presence of perennial snow and glaciers over the mountains is of great environmental and economic importance to South Asian countries and particularly for the Indian subcontinent. The perennial snow and ice at high altitudes, not only provides a great cooling effect to our Earth system (particularly to the Indian tropical region) but it also forms a constant source of water, comforting and influencing all life forms. The present volume entitled "Himalayan Snow and Glaciers — associated environmental problems, progress and prospects forms a good introduction to the subject and is informative.

I personally know the dynamism and keen interest of the author in the field of glaciology for last four decades. He used to come to IMD office for studying the effect of fluctuations of snow cover on the hydrology of Himalayan rivers. In fact, he persuaded me for a careful analysis of satellite APT pictures for the purpose. On his return from the fellowship in Norway, he gave several lectures on snow and glacier hydrology. He actively participated in the Inter-departmental Gangotri Glacier Study-cum-Training Expedition. He continued his interest in Himalayan hydrometeorology in active association with IMD. In spite of the deficiency of Indian data, Dr. Bahadur continued his interest in the field. He has been associated with the national and international programmes on Himalayan glaciology. In 1991 he launched an experimental campaign, calling it as Himalayan Experiment (HIMEX). At the instance of Department of Science & Technology, a three-day brain storming seminar was convened by him in 1995 at the Centre for Atmospheric Studies (CAS) at IIT, Delhi, which was attended by a large group of scientists and emphasized on data collection on snow, ice and other natural resources for sustainable development of Himalayan region. At the instance of Vigyan Prasar, recently I had an opportunity to review his book on *Indian Himalayas; an Integrated View*. He has painstakingly collected and collated the information on various facets of atmosphere, geosphere, hydrosphere and biosphere including associated natural hazards with mitigation strategies, institutional and infrastructural aspects along with socio-economic considerations and needed policy interventions.

I am sure that the present publication, giving information on Himalayan snow and glaciers along with historical background, international efforts for the study will arouse more interest in the study of apex snow and ice, effecting productivity of our nation. This assumes greater importance when global warming is having its adverse impact on this snowy environment. To me, it appears that there is an urgent need to have a national centre to monitor all aspects of Himalayan snow and glaciers feeding our northern rivers.

Dr. P.K. DAS
F.N.A., F.A.Sc.
Formerly Director General
India Meteorological Department
and WMO Consultant
A-59, Kailash Colony
New Delhi - 110 048

Preface

My interest in Himalayan snow and ice was kindled by the Union Ministries of Education, and Irrigation and Power by sponsoring me for the global post-doctoral fellowship under the Norwegian Agency for International Development (NORAD) for the study of snow and glacier hydrology. This provided a great opportunity to work under the guidance of Prof. Gunnar Østrem — a distinguished Glaciologist. During my stay in Norway, I interacted with several snow and ice experts from cold countries who all encouraged me for the study of high altitude Himalayan snow, ice and glaciers. On my return, I actively participated in Inter-departmental Gangotri Glacier Study Expedition, 1971 under the Indian National Committee on International Hydrological Decade (CSIR). During 1983-84, I was assigned to initiate and coordinate inter-institutional R&D investigations in the field of Himalayan Glaciology at the Department of Science and Technology. Meanwhile Director, National Institute of Hydrology, Roorkee offered me to prepare a document entitled, 'Snow and Glaciers and their Contribution to India's Water Resources' (February 1992). This publication prompted UNESCO to invite me to participate in the International Symposium on Snow and Glacier Hydrology and present a paper on 'The Himalaya — a Third Polar Region'. At this symposium, a plea was made to conduct Himalayan Experiment (HIMEX) for improvement of data-bases for the cold Himalayan region. In 1996, International Geosphere and Biosphere Progamme (IGBP) Secretariat invited me to participate in the workshop on Global Change Impacts on Mountain Hydrology and Ecology at the International Centre for Integrated Mountain Development (ICIMOD), Kathmandu and the UNESCO Conference on Eco-hydrology of High Mountain Areas. These opportunities further strengthened my conviction for an all-out effort to monitor Himalayan snow and ice, the apex renewable resource for the welfare of the nation.

Himalaya — the Abode of Eternal Snow (also known as '*Dev Bhoomi*', 'Nature's Paradise' and 'Abode of Peace and Tranquility') attracted me since my early days. I have tried to study the role of Himalayan snow, ice and glaciers for the hydrology of north Indian rivers; weather and climate and the great diversity of agro-eco-systems for last four decades. In this publication I have attempted to include various aspects of Himalayan snow, ice and glaciers.

In general, the Cryosphere (encompassing the perennial snow, glaciers and permafrost) and Atmosphere (the gaseous envelope which surrounds the earth) are highly inter-related through heat and moisture exchanges. The physical properties of snow, e.g. high reflectivity, high transmissivity, low vapour pressure and low conductivity, make it an important element of the climate system. Snowy Himalayan mountain chain, in the neighbourhood of hot tropical latitudes, has a great cooling effect on the environment and comforting influence on all forms of life. High altitude of the mountain with its large altitudinal spread provides a wide spectrum of micro-climates, making the Himalaya a unique mega-biodiversity zone on the planet. In fact, Himalaya is a world heritage zone of tremendous importance and a vast reservoir of natural and intellectual resources for the mankind. But Himalayan snow and ice are like blessings wrapped in a very difficult package as the high mountains are hardest approachable, rarefied and hazardous environment where even the strongest man finds it difficult to stay and work.

In the present publication, I have attempted to incorporate information on Himalayan snow and glacier fields; importance of Himalayan glaciers; international efforts for the study of Himalayan glaciers; perennial Himalayan rivers; soil erosion and sediment transport; environmental problems of high Himalayan region; monitoring for protection and conservation and utilization of modern technology. The text is illustrated with tables, figures and a few photographs collected from diverse sources giving references to the literature.

I am hopeful that the information given in the publication will be useful and encourage further investigations of apex snow and ice systems which largely govern our culture and agriculture.

Jagdish Bahadur

Acknowledgements

I express my most sincere obligation to the Secretary, Department of Science and Technology, Government of India for his constant encouragement and support for the difficult field of Himalayan Glaciology. Support provided by Dr N.K. Sehgal, Founder Director of Vigyan Prasar and the help for the preparation of the present text by all the staff members of Vigyan Prasar is gratefully acknowledged.

My sincere thanks are to my wife, Mrs. Saroj Bahadur for her constant support and maintaining a very conducive environment to make it possible to complete the publication. Contribution and willing cooperation of Shri J.R. Sharma for word processing the manuscript is worthy of appreciation.

The author thankfully acknowledges several suggestions for improvement in the presentation from my close colleagues Dr. V.B. Lal, Retd. Senior Scientist, CSIR and Shri N.K. Trivedi, Retd. Engineer-in-Chief, All India Radio and the sincere persuasion of Dr. B.D. Acharya, Adviser and Head, Earth System Science Division of Department of Science and Technology for the publication. My very grateful thanks are to Professor Harjit Singh, Head, Centre for the Study of Regional Development, Jawaharlal Nehru University, New Delhi for taking extra pains to review the text and for suggestions to include more technical information and photographs to enhance the value of publication.

Contents

List of Tables

List of Figures

Plates Coverage of Snow and Ice Conditions in Himalayan Region

1

Introduction

1.1 The Himalayas

The king of mountains known as 'Nagadhiraj' is the world's most active land — atmospheric mountain system. It is also the world's youngest, highest and most sensitive and extensive mountain system having 14 peaks over 8000 m and hundreds over 7000 m and 530 peaks above 6000 m high. Due to heavy snow precipitation at high mountains, its extensive snow and glacier ice fields rival those of polar regions. Therefore, it is sometimes called as a third pole on the planet earth.

The "Himalaya" is a Sanskrit word, literally means 'storehouse of snow and ice', and is more commonly referred to as the abode of eternal snows. It always holds an indescribable charm, inspiration and spiritual significance to the millions in the world.

Geological evidence exists which shows that the Himalaya have been formed by violent crumpling of the earth's crust along the southern margin of the great table — land of Central Asia, i.e. Tibetan Plateau (Chinese call it as Qinghai-Xizang or simply Qing-Zang Plateau). Uplift of Himalayas was a gradual process protracted over a very long period. This process had a very marked effect upon the scenery, the topography and the river systems.

It is now believed that during the period 10 to 50 million years ago, the subcontinent of India riding over its own tectonic plate, collided with the rest of the Euro-Asian land mass. The pressure squeezed the earth crust, forcing it to prop further inland. The uplift of Himalayas took place in three stages: (i) Early Miocene (70-21 million years); (ii) Late Miocene (17-7 million years) and (iii) (1.6 million years to the present). The Himalayan uplift estimates vary from 2 to 12 mm per year while the Tibetan Plateau has shown an average uplift rate of about 5 mm per year for last 10,000 years (Stone, 1992). The uplift is still continuing and is manifested in the form of regular earthquakes.

During the slow process of uplift, folding and faulting, the rivers were able to keep largely to their original courses. The erosive power of these streams was increased with the increase of their bed slopes. Thus, the rivers are formed cutting through the main chains of ranges in deep transverse gorges after flowing long distances parallel to the trend of chain. Most of these rivers drain not only the southern slopes of the Himalayan mountains, but to a large extent, the northern Tibetan slopes as well. In fact, the watershed of the chain is not along the line of the highest peaks but quite a distance north of it.

Generally, Himalayan mountain system is divided into three longitudinal zones with marked orographic features.

The Great Himalayan ranges, rising above the snowline with average elevation of 6100 m. The range is composed of high grade metamorphic and old granite rocks.

The Lesser Himalayas ranges south of the Great Himalayas with average height of 2600-4600 m, predominantly composed of mid to late Proterozoic sedimentary rocks overlain by early granitic and metamorphic rocks.

The Outer Himalayas with average elevation from 1000 to 1300 m (Shivalik ranges), consisting of sedimentary deposits laid down by rivers in the channel and flood plains.

In the outer valleys of Himalayas, the side slopes are so steep that landslides frequently occur. Simultaneously, the streams are active in cutting down the riverbeds reaching their base levels. All these streams are very turbulent and are perennial due to regular feeding from snow and glacier melt drainage from the high mountains.

1.2 Snow

Snow is the solid form of water which grows while floating, rising or falling in the free air of the atmosphere. A snow crystal is intriguingly beautiful and has a sand particle at its centre. Snow crystals have infinite variety of forms and shapes. More than 6000 different forms of crystals have been photographed but this listing is still considered incomplete (Upadhaya, 1995). Three basic forms of snow crystals are hexagonal prismatic column, thin hexagonal plate and branching star shaped form, sometimes called dendritic. Various forms of snow crystals are associated with the temperature of the surroundings. Grain size of the snow crystals refer to the size (length) of the majority of crystals in a given sample. It could range from less than 0.5 mm to more than 4 mm. Freshly fallen snow have a density as low as 0.01 g/cc, i.e. a mixture of 1% moisture and 99% of air. Its textural and physical property continuously change as the crystals deform and density changes occur. The density range of snow forms of various types (Paterson, 1969) are new snow at low temperatures in calm (0.01 – 0.03 g/cc); immediately after falling in calm (0.05 – 0.07 g/cc); damp new snow (0.1 – 0.2 g/cc); settled snow (0.2 to 0.3 g/cc); depth hoar (0.2 to 0.3 g/cc); wind packed snow (0.35 to 0.4 g/cc); firn (0.4 to 0.85 g/cc); very wet snow and firn (0.7 to 0.8 g/cc) and glacier ice (0.85 to 0.91 g/cc). Probably, the greatest importance of snow and ice is biological just as snow provides insulation which permits plants and a small animals to survive the severity of winter, so the cover of ice over a lake or river prevents the loss of too great a quantity of heat protecting the aquatic flora and fauna from cold injuries.

1.3 Glaciers — Living Ice

In simplest term, a glacier can be defined as a body of natural land borne ice that flows. Some call it as a river of ice. A glacier constantly changes its shape and form to adjust to the changing environmental conditions. It is, therefore, called as living ice. Glaciers occur in those parts of the earth where the rate of snow precipitation is greater than the rate of melting of snow. The ideal locations are the polar regions and the high mountain ranges. The glaciers are classified depending on morphological (area – altitude) characteristics; thermal conditions and their dynamic behaviour, they move due to internal flow of ice and slippage over bed rock. Warm glaciers move more rapidly. The increased velocity results in extending flow and the glacier advances and thins. Decrease in velocity results in compressive flow and the glacier shrinks and thickens. Due to kinematic wave, increased discharge occurs through a glacier. Due to surge, a large and rapid increase in basal slip occurs and the glaciers move very rapidly creating hazardous situation for the downstream inhabitants.

1.4 Ice Ages

The planet earth has experienced recurring ice ages (i.e. the spread of regional glacier ice) for the past few million years. Ice ages are almost certainly trigger for slow "wobbles" in the earth's axis and in its orbit around the sun. These wobbles affect the total amount of energy the planet receives from the sun and in particular its geographical distribution. During an ice age, global temperatures fall by about 5°C and the snow and ice mass increase in areal extent and volume due to increased cooling. Under the reduced temperature conditions, larger snow would accumulate in mountains and glaciers would advance and the continental ice sheets return creating deep-freeze environmental conditions. There are several hypotheses for the occurrence of an ice age. They include wandering of poles, reduction in carbon dioxide in atmosphere, earth's movements resulting in profound uplifts, eruption of volcanic causes, drift of continents, changes in the ocean water circulation, variation in solar radiation but the definite cause for occurrence of an ice age is not clearly known. Glacial deposits and landforms show several short glacial stages separated by longer periods of warm climate (inter-glacial stages) when the plants and animals thrive. The glacial and inter-glacial periods have had major influence on a species extinction and evolution of flora and fauna.

For 250 million years, throughout the age of dinosaurs (extinction about 70 million years before present), the earth was warm and wet, almost tropical. It was around 40 million years ago, the planet began to cool (Paterson, 1993). Around this time, India slammed into Asia and new habitats were created due to upheaval and uplift of Himalayas that helped trigger the ice age. The vast high plateau of Tibet may have played a large part in reducing earth's temperature. Some geochemists now suggest that the rising plateau, forced up from the sea floor as India crashed into Asia, was responsible for reducing the concentration of carbon dioxide in the atmosphere. The gas traps heat radiated from the earth's surface, so as the level of carbon dioxide dropped, the earth radiated more energy from the sun and began to cool. This is supported by isotopic ratios of strontium ($^{87}Sr/^{86}Sr$), Osmium ($^{187}Os/^{186}Os$) and oxygen ($^{18}O/^{16}O$) of the ocean sediment cores.

1.5 Glaciation

Unlike Europe and North America, Asia was not covered by a continental ice sheet during Quaternary (geological period of last two million years ago). It is represented by local accumulation of glacial (Pleistocene, i.e. the geological period from 2 million years ago to 10,000 years before present) and post glacial (i.e. Holocene — the geological period from 10,000 years before present to the present), which continue without change of fauna. During Pleistocene, the glaciers occupied 30 per cent of the total area of the earth's land surface as against 10 per cent at the present (Williams, 1986). The Quaternary appears to be an artificial division of time to separate pre-human and post-human sedimentation.

The most recent glaciation on earth, reached its maximum advance about 20,000 years ago. Moving downhill at up to several metres a day, the ice — nearly 1600 metres deep in places, scrapped the earth like a giant bulldozer, pulverizing the rock before it (Bode, 1990). As the earth warmed, glacier retreated up the steep valleys and the melt water at the leading edge of the ice put down thick layers of loess in the low land. Borne by centuries of winds, the "China Dust" settled in deep deposits in the Manturian and North China plains, helping to create some of the richest farm lands in the world.

In Europe, ice age winds sweep down the tundra resettled

glacial silts in a blanket 85 m deep, to create agricultural plain that is stretched from Germany to Ukraine. The same dust-like mantle is 45 m deep across the "bread basket" of Central Canada and the United States. But nowhere are the loess drifts as deep as China, i.e. 400 m in places. Ice ages gave us clue about the appearance of man at different places. It is now widely accepted that the early man (Homoerectus) first arose in Africa between 1 million to 2 million years ago. Scientists unearthed bones buried in China loess, belonging to Lantian man, who lived about six hundred thousand years ago. Near Beijing, they found bones belonging to the Peking man who lived between five hundred thousand to two hundred fifty thousand years ago. Modern man (Homosapiens) did not appear in Asia until fifty thousand years ago.

Himalayan glaciation has been a controversial issue for a long time. Throughout the Himlayan region, extensive evidence of past glaciation is provided by well developed moraines and valley fills that exceed several tens of metres in thickness. Although relative chronologies have been established for several parts of the Himalayas, the timing of glaciation is poorly understood because of paucity of suitable dating material. Researchers have often assumed that Himalayan glaciations were synchronous with global ice-volume maxima implying that the climatic change in Himalaya is tied to temperature cycles in the northern mid-latitudes. However, new dates based on ^{14}C, Thermo-Luminescence, Optically Stimulated Luminescence (OSL), U/Th in collaboration with $\delta^{18}O$ of ice cores do not support this assumption (Gillespie, A & Molnar, P., 1995; Sharma, M.C. & Owen, L.A., 1996; Owen, L.A. *et. al.*, 1997), suggesting that the influences on climatic change in Himalayas are more complex than previously thought. A review of the currently available dating evidence (Benn, D.I. & Owen, L.A., 1998) for the timing of Himalayan glaciations and paleoclimatic records from adjacent regions, examined the possible implications for climate dynamics in the region. Himalayan glacier fluctuations are controlled by variations in both the South Asian monsoon and the mid-latitude westerlies. It is concluded that the glaciers in this region were less extensive during global glacial maxima due to increased continentality of the climate of High Asia.

1.6 A Fortunate but Fragile Heritage

Himalayan's sense of unspoiled beauty and splendour is dazzling but unfortunately misleading. Himalayan snow and ice are blessings wrapped in a very difficult package as they are the hardest approachable and one has to work in a rarefied high altitude and hazardous environment. There are both benevolent and malevolent effects to Indian population. Among the fortunate and benevolent effects are enormous cooling of otherwise hot environment, gift of Indian monsoon, perennial water supply for irrigation and power generation, moderation of floods and droughts in the Indo-Gangetic plains. Among the malevolent effects are cloud bursts and disastrous glacial phenomena like intense freeze — thaw cycle resulting in strong soil erosion, snow and glacier avalanches, glacial mud flows, outbursts of glacier dammed lakes, land slips and flash floods besides frequent earthquakes. Without Himalayas, the Indian landmass would have been a hot arid subcontinent with difficult living conditions. Model studies conducted at National Center for Atmospheric Research, Colorado, USA shows that with no mountains, the temperature over southern Asia would have been 12 degrees centigrade higher and there would have been no Indian monsoon — one of the most spectacular atmospheric phenomena in the world.

Himalayan ecological system is not as fragile as it is sometimes made out. It is a very dynamic system having inter-related atmospheric, geo-spheric and bio-spheric aspects. It can be said that the system has a fluctuating stability, having the capacity to readjust itself to the enormous stresses created by human activity such as creation of large man-made reservoirs. What is needed is to better understand the spatial and temporal changes in various physical processes affecting land and the life on it in a domain of ecological system (Gerrard, 1990) and this needs its monitoring.

2

Himalayan Snow and Glacier Fields

Himalayan snow and glacier fields have all along been actively and constantly modifying the Himalayan terrain. A brief account of historical developments and observations made by several explorers have been listed chronologically. Separate units, divisions and organizations under various departments of Government of India, to undertake studies on snow and ice are included. Extent of snowfields, characteristics of Himalayan glaciers, and influence on weather and climate are outlined.

2.1 Historical Background

Kennith Hewitt comments that the European science in the mountain of Central Asia is virtually absent from today's histories of geomorphology. It is more than accidental neglect. Before 19th century, the works of Chinese and Arab geographers were superior to their European counterparts (Hewitt, 1989). Some of the major highlights of scientific contributions are briefly described and chronologically listed:

1659: Francois Bernier wrote of 'grand Tibet' as the hopeless country of mountainous deserts.

1842: Vigne contributed to recurring theme of the glacier dammed lakes and outburst floods. He visited the snout of large Chogo Lungma glacier stating that he had never seen any spectacle of the same nature so truly grand as the debouchure of water from beneath this glacier.

1853: Strachey gives a good picture of Indus from Mansarovar to Gilgit and comments:
"the immense extent of existing alluvium and the uniformity of its maximum elevation, led me to infer that it must have been under a general sea covering the whole country —and not by lakes, much less rivers — The gigantic ravines in the ancient plateaux and out of proportion to the contemptible rivulets that now creep along their bottoms, wandering overbeds a thousand times too large for them.
He also visited the terminus of Siachen glacier and documented three damburst floods.

1854: Cunningham was the first to infer substantial climatic change in Central Asia, believing the region was more humid and milder in the past and the vast lake systems describing fresh water mollusks in sediments of now saline lakes.

1860: Crodwin Austen attributed much of the extensive superficial deposits to far larger glacier cover.

1861: Three brothers (Hermann 1826-82, Adolf 1829-57 and Robert Von Schagintweit 1833-85) travelled extensively covering mountain from Assam to Afghanistan and were first to demonstrate the lower altitudes reached by Karakoram glaciers. They attempted to define the interrelated patterns of climate, vegetation and orography. They gave considerable attention the extent of contemporary glaciers and snow covers and pointed out to topoclimate effects how the region differed from Alps — the mountains of lesser relief. They regarded the glacier fluctuation in Asia as reflecting local rather than global climatic influences and to have been of lesser extent and more recent origin than those of the Alps.

1891: Tanner described the 'havoc' committed by snow avalanches in 1879 in many valleys' of Kashmir and Gilgit.

1892: The "antecedence" hypothesis of Himalayan rivers

was applied as it was accepted that the gorges of the rivers across the Himalayas were cut by the rivers not by a marine inundations and during the rise of the mountains.

1894: Conway was the first to stress the exceptional scale and frequency of avalanches at high altitudes.

1896: Dierner reviewing the problem of the ice age in Himalayas concluded evidence showed a maximum descent of glaciers to 2100 m.

1904: French geographer C. Robert presented mudflow as a dominant feature of the Karakoram environment.

1906: Karl Destriech argued that the transverse Himalayan courses of the Indus and other rivers are "antecent and epigenetic".

1907: A sketch of the Geography and Geology of the Himalayan Mountains and Tibet (1907) by Col. S.G. Burrard (Supdt., Trigonometrical Surveys) and H.H. Hyden (Supdt., Geological Survey of India) whose preface states:

One subject has fallen naturally into 4 parts as follows:

Part I – The high peaks of Asia.
Part II – The principal mountain ranges of Asia
Part III – The rivers of the Himalayas and Tibet.
Part IV – The geology of the Himalayas.

1913: Kenneth Mason (1887-1970) began his lifelong association with the Karakoram and became especially involved in the well-established problem of those glaciers that interfered with rivers and created outburst floods.

1917-22: Hedin, the founding figure of modern physical geography and geomorphology propounded one of the major hypothesis of landform development — notion of expanding river systems in Central Asia, related to climatic change and orogeney, with glaciation and fluctuating lake levels playing important role.

1918-34: Gotto Dainelli edited several volumes of Geographical and Geological Results. Scientific understanding of

Karakoram are most evident from the following:

I. History of Exploration (1934) 430 p.

II. Historical Geology (1933-34) Pt. I 458 & p. II 1099 p.

III. Ice Age Studies Pt. I 651 Pt. II (Illustrations).

IV. The contemporary Physical Environment (1928) 479p. with Marineli. Much later, he wrote a more popular work on Karakoram exploration in 1958.

1933-34 : Dr. Terra (1900-1981) agrees with Dainelli estimate of 8000 ft. of uplift in the Pir Panjal Range, to the South, since the second half of the Pleistocene.

1930-39 : Wadia's path breaking work dealing with exceptional central thickness, tectonic activity and rates of uplift in what he called the North West Himalayan Syntaxis. The work was resumed after Second World War (Ganser, 1964).

The remoteness, great scale, extremes of climate, altitude and ruggedness of the region must not be forgotten. They have absorbed much more effort than scientific tasks. A great deal of what we have in reconnaissance rather than sustained and systematic observation.

Scientific interest in snow and glaciers of the Himalayan began observations on the snow line or 'line of perpetual snows' in the 1840s. Blandford wrote his first 'Note on glaciers in Hindustan' in 1873. Subsequently he made additional remarks on Himalayan Glaciers in 1877 and also a 'Note on Age and Glaciers of the Himalayas' in 1891. Glaciology was thus fairly launched in India by nineteenth century. The pattern of occasional and incidental observation of Himalayan Glaciers was basically altered in 1907-10 when the Geological Survey of India (GSI) made a concerted effort at systematized glacier observations as a part of International Programme to study glaciers.

A second effort for glacier observations was organized during International Geophysical Year (1957-58) when many glaciers were revisited to study their fluctuations. In 1945 a paper on the role of glaciers and snow hydrology of Punjab rivers was presented at the Annual Research Committee meeting of Central Board of Irrigation and Power (Kanwar Sain, 1946). Following this Dr. J.E. Church, formerly President of the International Commission on Snow and Glaciers was invited to India, by the Prime Minister Pt. Jawaharlal Nehru, immediately as the country became independent, to initiate snow surveys and stream flow forecasting in the Himalayas and train the necessary personnel in the specialized field of high altitude science (Dhir and Singh, 1956). Unfortunately, the work was not pursued with vigour due to hardships involved and lack of adequate financial and institutional support. During 1969, the Indian National Committee on International Hydrological Programme (1965-74) under CSIR (Hydcom), appointed another high level committee on Snow, Ice and Glaciers. Under the guidance of this High Level Committee, Inter-Departmental Glaciological Expeditions were organized to various Himalayan Glaciers. The studies included hydro-meteorological observations, glacier dynamics, mass balance, geomorphological and isotopic investigations, ice drilling and experiments for artificial melting of snow and ice. Geological Survey of India undertook the preparation of glacier inventory on 1:50,000 scale with cooperation from Survey of India. A UNESCO regional workshop on snow, ice and avalanches was organized at the Snow and Avalanche Study Establishment (SASE), Manali in 1977 followed by a training seminar in 1978.

Several Ministries under the Government of India established separate units/divisions and organisation to undertake studies of Himalayan snow and ice. Predominant organizations are:

Snow and Avalanche Study Establishment (SASE), Manali

SASE was established in late sixties by the Ministry of Defence with the sole objective to study snow cover and meteorology, snow-mechanics and avalanche dynamics,.snow physics, artificial triggering of avalanches, available technology and control structures along with field trials with a view to facilitate movement over high altitude roads constructed by Border Road Organisation.

Geological Survey of India (GSI), Calcutta

Systematic glaciological studies in GSI was initiated in 1974 with the creation of Glaciology Division. Present studies cover glacier inventory; glacier regimen, glacier and snow hydrology, suspended sediment transport, seasonal snow cover assessment, ice flow movement and other specialized studies.

India Meteorological Department (IMD), New Delhi

India Meteorological Department (IMD) established a Glaciological Study and Research Unit during 1972-73 for taking active participation in national glacier expeditions. Activities include meteorological observations, such as precipitation, temperature, wind speed and direction, short wave, long wave radiation for energy budget and mass balance studies on Himalayan glaciers.

Central Water Commission (CWC), New Delhi

Central Water Commission (CWC) was the first organization to initiate snow surveys for stream flow studies. In 1985, CWC established a Snow Hydrology Unit at Shimla for flood forecasting of Upper Yamuna Catchment. Recently, National Institute of Hydrology (NIH), Roorkee has been actively involved for mountain hydrology of the snowy and glaciated Himalayas. Being the nodal organization for water resources exploration and management, CWC has yet to establish a suitable network of observation stations for evalauation of meltwater contributions from Himalayan snow and ice.

Survey of India (SoI), Dehradun

Survey of India (SoI) has always been associated for topographical surveys and preparation of large scale maps of glaciers and its surroundings including snow and moraine features, rock fall, ice walls, crevasses, snouts, dead moraines, caves, glacier movements. These studies are coordinated by the Geodetic and Research Branch of Survey of India.

Wadia Institute of Himalayan Geology (WIHG), Dehradun

Wadia Institute of Himalayan Geology (WIHG), Dehradun under the Department of Science and Technology has played a lead role since 1986 for multidisciplinary scientific expeditions under the Himalayan Glaciology Programme. The Institute established a semi-permanent field research station near Dokriani Glacier at an altitude of 3750 m in 1992. Field laboratory facilities for sedimentological, melt water discharge and meteorological studies. Continuous monitoring of glacier for the whole ablation period of six months has been achieved for the first time. The Institute has acquired sophisticated modern equipment such as Ground Penetrating Radar (GPR), Particle Size Analyser, Automatic Weather Station (AWS) and Electronic Distance Meter (EDM) to produce high quality glacier data. Other testing facilities at the Institute are available for research purpose.

Physical Research Laboratory (PRL), Ahmedabad

Physical Research Laboratory (PRL), Ahmedabad under the Department of Space has been studying snow and ice problems using isotopic studies. Glacier dynamics (using ^{32}Bi, ^{210}Pb and ^{137}Cs) and climatic variations (^{18}O, ^{2}H) and chemical composition of snow and ice. Studies for deep ice core from glacier ice for climatic variations and chemical composition have been identified as the future thrust area.

National Remote Sensing Agency (NRSA), Hyderabad

National Remote Sensing Agency (NRSA), Hyderabad, under Department of Space has been providing seasonal and short-term snow melt run-off forecasts for various Himalayan river basins on an operational basis for last two decades. Himalayan Snow cover Information System (HIMSIS) is being implemented on a work station with PC systems as clients having necessary state-of-the-art technology, hardware and commercially available as well as in-house developed software. Digital Terrain Models (DTMs) have already been designed with Ingress RDBMS core for 23 basins since 1987 for Water Resource Management.

Space Application Centre (SAC), Ahmedabad

Space Application Centre (SAC), Ahmedabad under the Department of Space has initiated remote sensing studies from satellite imageries, the thrust area identified by the organization are snow mapping and snow hydrology; inventory of glaciers and mass balance of glaciers. Hazards from glacier dammed lakes of Sikkim are being monitored.

The first national symposium on Seasonal Snow Cover was organized by SASE, Manali at Vigyan Bhawan, New Delhi in 1983. A national meet on Himalayan Glaciology was organized by the Department of Science and Technology in 1989. In 1994, SASE organized an international symposium on 'Snow and Related Manifestations'. In 1995, a three-day seminar was supported by the Department of Science and Technology at the Centre for Atmospheric Sciences, Indian Institute of Technology. Another major activity was the symposium on Snow, Ice and Glacier organized by Geological Survey of India in 1999 at Lucknow which was followed by a brain-storming session on Mathematical Modelling for Glacier Studies organized by the National Academy of Sciences, India at Allahabad under Himalayan Glaciological Programme of DST.

Time has come now to organize a separate institute for Himalayan Snow and Glacier studies with closer inter-departmental collaboration and a clear mandate. We have to strive to seek answers to the following :

(i) What is the total reserve of fresh water locked in perennial snow and ice ?
(ii) What is the normal accumulation of seasonal snow cover and how its spatial and temporal variability affects stream flow of different Himalayan rivers ?

(iii) What are the effects of Himalayan snow and ice on local climate and regional monsoon circulation?
(iv) What are the possibilities to use glaciological research data for modelling climate and climate changes?
(v) How to harness glacial melt for managing droughts and floods ?
(vi) What are the technological gaps requiring upgrading of infrastructure/development of equipments and instruments for the study?

2.2 Extent of Himalayan Snow Fields

A sufficient area of the high mountain system lie higher than the snow line (isotherm 0°C) and in this area, snowfalls round the year. It is estimated that 10 to 20 per cent of the local surface area is covered by glaciers while an additional area ranging from 30 to 40 per cent is overlain by seasonal snow cover.

Snow cover in the Himalayas occurs and exists depending on the terrain and climatic conditions of the region. The snow cover could be temporary, seasonal or permanent. The temporary snow cover stays for a few days and then melts away. It occurs at lower altitudes during winter and even sometimes at higher altitudes during summer. Snow cover that is formed over weeks or months by consecutive snow falls and which melts away gradually during the following summer, is termed as seasonal snow cover. Above a certain altitude and in certain situations, some amount of snow is carried over to the next winter season without melting away during summer, and this turns into firn and ultimately ice adding to the permanent snowfields and glaciers. It is the seasonal snow pack that contributes to the water resources during the lean summer months (April, May, June) and is thus most important. The duration for which seasonal snow pack stays is variable with respect to basin location and also from year to year.

No consolidated information is available for the extent of snow cover from year to year in Indian Himalayas. It may be interesting to note that the inter-annual variability of Himalayan snow cover is substantial (Bamzai & Kinter, 1997) as shown below :

Table 2.2.1: Interannual Variability of Himalayan Snow Cover (10^6 km^2)*

Winter season		*Spring season*		*Summer season*	
Mean	*S.D.*	*Mean*	*S.D.*	*Mean*	*S.D.*
2.20	1.25	1.11	0.73	0.48	0.43

Note : S.D. stands for Standard Deviation.
**Source* : Bamzar, A.S. and Klinter, J.L., 1997.

In a recent investigation, dynamic characteristics of snow cover in Western China (Peiji, 1994) are reported. The study presents a detailed and accurate spatial pattern, seasonal cycle, and year to year variability of snow cover for 30 year period, 1957-87 by using the best information from micro-wave derived snow depth charts, employing data from Scanning Micro-wave Multi-channel Radiometer (SMMR), operational NOAA digitized weekly snow cover extent charts, (Fig. 2.3.1) short-wave Defence Meteorological Satellite Programme (DMSP) imagery, in conjunction with daily snow depth records for 175 ground climate stations.

The results show that the snow season begins in mid-September over the Tibetan Plateau. From mid-October to early November, snow cover development is rapid. It increases to late November — early February. The peak with maximum snow cover occurs in January showing snow cover area of about 1.5 million sq. kms. This is followed by a slow decline until June. During the growth progression of snow cover, the increase in a real extent leads to snow volume increase, whereas in the decay phase, decrease in snow volume is followed by decrease in snow area. The seasonal progress of snow cover in lowlands in Western China, derived from station data is different from that in Tibet in two ways. In lowlands, the snow begins one month later and ends two months earlier. The winter, spring and autumn are represented by 45.2 per cent, 28 per cent and 21.2 per cent of the snow volume respectively over the Tibetan Plateau. The larger year to year variability of snow cover is most striking feature. The variance in snow volume is more noticeable than that in snow cover area. The ratio between the highest annual volume and the lowest is 1.91 for Tibetan Plateau and 1.39 for Western China while the area ratio is 1.59 and 1.27 respectively.

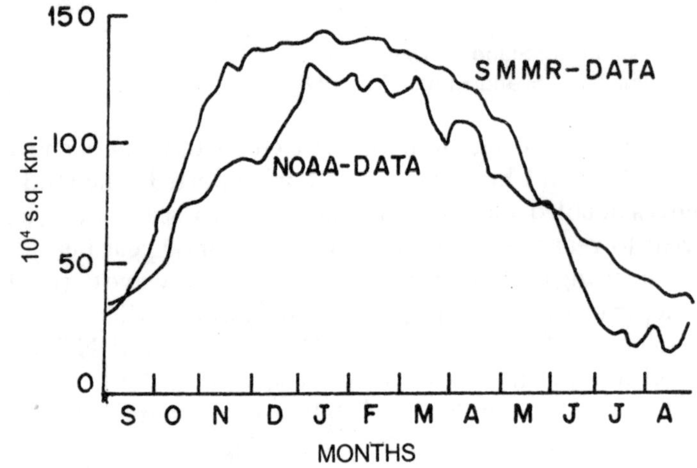

Fig. 2.3.1: Normal Annual Cycle of Himalayan Snowcover.

The Himalayan snow cover is one of the most significant parameter for the feedback mechanisms controlling the activity of Indian summer monsoons.

2.3 Extent of Himalayan Glacier Fields

Glaciers of high Asia cover an area of about 50 per cent of all glaciers outside the polar region. Their areal extent is 33 times the glacier area occupied by the European Alps. Table 2.3.1 gives an approximate distribution of permanent snow and ice in Himalayan mountains (Upadhaya, 1995) having a significant cooling effect on their immediate neighbourhood, the regional and the global environment.

In 1984, Chinese scientists have published that the total glacier area in the Himalayan region is 94,554 sq. kms. (Jijun and Shuying, 1984). The percentage area is shown in brackets in Fig. 2.3.2 for various mountain systems. It is also claimed that 49.8 per cent of these glaciers lie in Chinese territory as shown in Table 2.3.2:

Table 2.3.1: Distribution of permanent snow and ice in Himalayan Mountains*

Sub-region	*Extent of permanent snow and ice*	
	Volume (km^3)	*Surface area (km^2)*
Hindukush	9.30	6,200
Karakoram	2,180	15,670
Himalayas	5,000	43,000
Tibet	4,820	32,150
Total	**12,930**	**97,020**

**Source* : Upadhyay, D.S., 1995.

Table 2.3.2: Glaciers on the Qinghai-Xizang Plateau and its Marginal Mountain Ranges*

Sl. No.	*Mountain Range*	*Glacier Area*		*Glacier Area within Chinese Territory (km^2)*
		Area (km^2)	*Percentage*	
1.	Himalayas	29,685	31	11,055
2.	Karakoram	17,835	19	3,265
3.	Kunlun Shan	11,639	12	11,639
4.	Pamirs	10,304	11	2,263
5.	Nyenchin Tangla Shan	7,536	8	7,536
6.	Hindukush	6,200	7	0
7.	Chang Tang	3,566	4	3,566
8.	Gangdise Shan	2,188	2	2,188
9.	Tanggula Shan	2,082	2	2,082
10.	Qilian Shan	2,063	2	2,063
11.	Heng Tuan Shan	1,456	2	1,456
	Total	**94,554**	**100**	**47,113**

**Source* : Jigun, L, Shuying, X 1984.

The glaciers on different ranges exist in the altitude range varying from 4,000 to 6,200 m and the maximum altitude range is covered by the Himalayan Glaciers due to high precipitation on higher Himalayan ranges.

Out of several thousands of glaciers some important glaciers (Bahadur, 1972) are listed in Table 2.3.3 along with their approximate lengths and the region The major clusters of glaciers occur in and around major Himalayan mountain peaks:

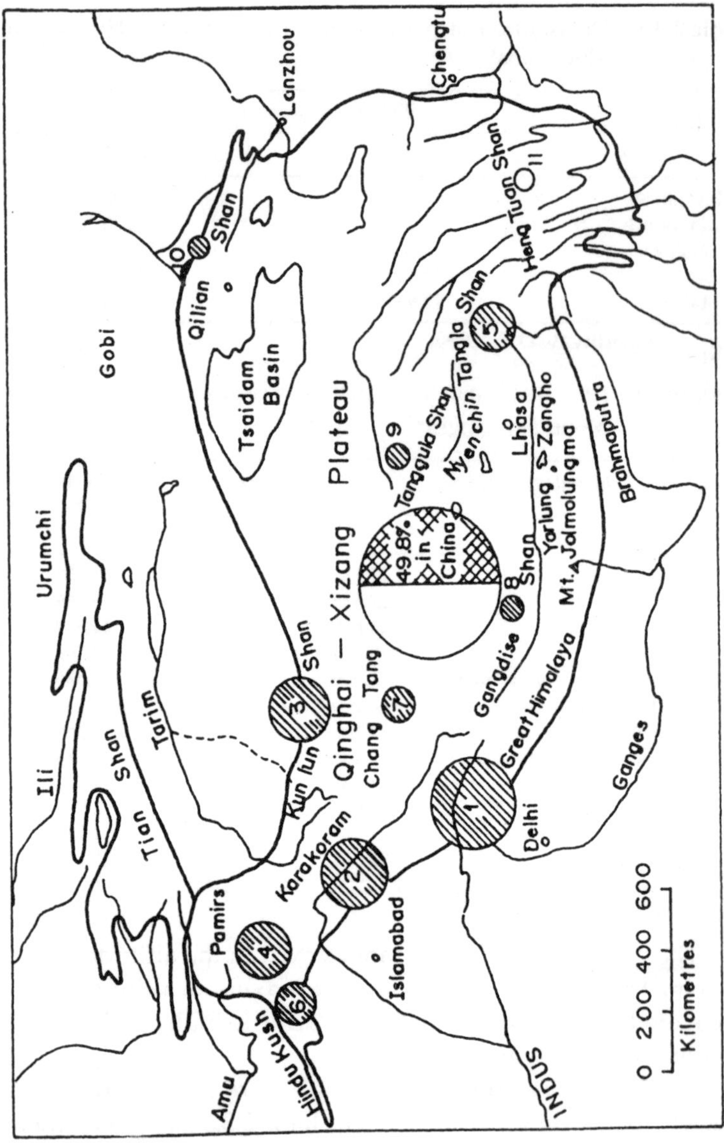

Fig. 2.3.2: Distribution of Himalayan Glaciers including Tibetan Plateau

the Nanga Parbat; the Nandadevi; the Dhaulagiri; the Everest-Makalu; the Kanchenjunga; the Kula Kangri and the Namche Barwa; the K-2 region. Altitudes of glacier termini in the Himalaya were plotted by Fujii (1977) as shown in Fig. 2.3.3. The termini altitudes are low in the western Himalaya as they are located at higher latitudes and are largely governed by the winter precipitation as compared to those in Central and Eastern Himalayan which lie in monsoonal regime. Distinct differences can be seen between the termini altitude on the southern slope and those on the northern slope: lower altitudes of the termini on the south slope of the Himalayas reflect higher accumulation due to monsoon.

Table 2.3.3: Some important Glaciers in Himalayas*

Region	*Name of the glacier*	*Approximate length (kms.)*
Eastern Himalayas (Sikkim)	Zemu	26
	Kanchenjunga	16
Central Himalayas (Kumaon)	Milam	19
	Kedarnath	15
	Gangotri	26
	Kosa	11
Western Himalayas (Punjab & Kashmir)	Rupal	16
	Diyamir	11
	Sonapani	11
	Rundum	19
	Punmah	27
	RIMO	40
	Chong Kumdon	19
	Nivapin	NA
North Western Himalayas (Karakoram)	Biafo	63
	Hispar	61
	Baltoo	58
	Gasherbrum	39
	Chogo Lungma	39
	Siachin	72
	Batura	58

**Source*: Bahadur, J. (1972).

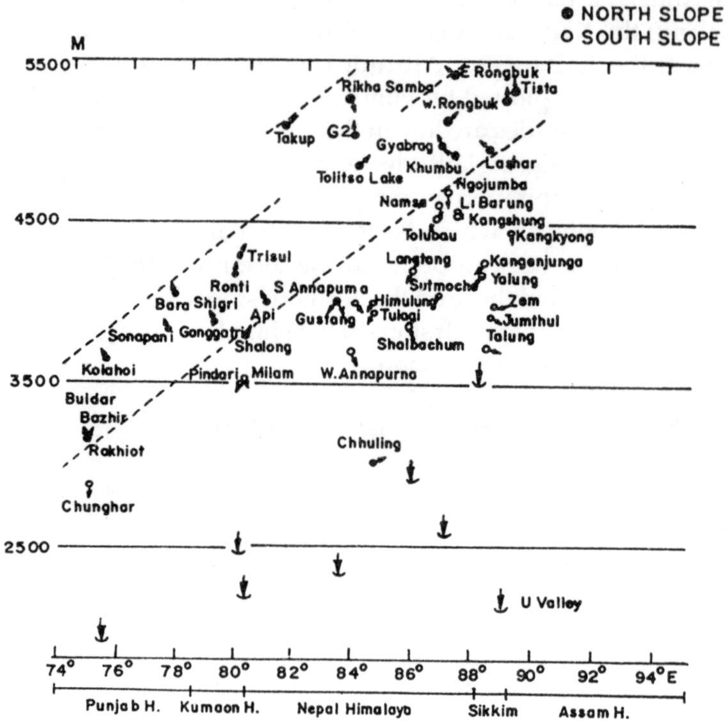

Fig. 2.3.3: Altitude of glacier termini in the Himalayas

The snowline, or the lowest limit of perpetual snow on the southern slopes of Himalayas, varies from 4300 m in eastern Himalayas to 5800 m in the western. On the Tibetan side, the snow line is about 900 m higher owing to desiccation caused by the absence of moisture-bearing winds. However, the permanent snowline on glaciers in western Himalayas is lower (4250 m) due to difference in latitudes and higher winter precipitation as compared to that in eastern Himalayas (4570 m).

2.4 Characteristics of Himalayan Glaciers

In general, Himalayan glaciers cover large altitude range as compared to glaciers in other mountain ranges located in non-polar regions. They cover an elevation range from a few hundred to more than 2000 m. Due to their large altitude range, they pass through different climatic and wind zones, which influences the accumulation and ablation of snow.

Himalayan glaciers differ from those in the Alps and other high latitude glaciers in the following features:

(i) Debris cover over the glacier surface is more dense particularly in ablation zone. Dust spread is also more as a consequence of mechanical destruction of rocks and boulders. In addition, there is a significant quantity of biomass and bacterial spread over the Himalayan glacier surface. Hence, these glaciers have blackish or brownish or un-shining look as compared to the mountain glacier of higher latitude. This reduces surface albedo to 10 to 20 per cent or even less on many glaciers.

(ii) Summer temperatures over the Himalayan glacier is higher particularly on southern slopes. It is usually above 10°C. Occasionally, temperatures of 18-20°C are also experienced at lower altitudes, resulting in more liquid water content of the snow and firn layers.

Both the above features tend to enhance melt rate resulting into large water yield to river run-off.

2.5 Influence on Weather and Climate

Glaciers, being the progeny of climate, interact with it and also its change at different scales.

The climate on earth–ocean–atmosphere system has fluctuated between temperature limits up to 15°C for hundreds of millions of years (Sharp, 1988). General trends in global climatic change for various time-scales ranging from decades to hundreds of million years have been the topic of studies by several earth scientists. The paleoclimatic situations have helped them to reconstruct climatic change from records of instrumental data, historical information, marine plankton, several sea level terraces, variations in isotopic concentrations in deep sea sediments and glacial ice cores.

In general, the climatic conditions that determine the regime and annual value of glacier melting and evaporation exhibit a great diversity in the region, The differences in circulation processes in the NW and SE parts of Himalayan region produce opposite annual regimes of cloudiness, precipitation and air humidity.

The summer temperature of air-glacier interface on the leeward slopes drops –4 to –2°C as the elevation increases. The rise of temperature southwards is almost neutralised by the growing monsoon cloudiness (Kotlyakov and Lebedeva, 1998) For example, at an elevation of 5000 m in the greater Himalayas, where cloudiness reaches seven to eight point, the summer temperature is barely around 3°C. During the same period in the Hindu Kush, at a point 10° farther to the north latitude-wise, the temperature is only 3-4°C Lower. Here the weather is dry and cloudless, and this is associated with the extensive Asian thermal low. Because of the cloudiness, the pressure of water vapour increases in the direction from NW to SW from three to four GPa to seven GPa and more.

Winter temperature distribution is governed by the latitudinal intensification of solar radiation influx to the south due to the increase of sun elevation and of day duration. At an altitude of 5000 m in the Hindu Kush, the winter temperature of the air above glacier is equal to –16°C whereas the greater Himalaya it is only –6°C. The difference is further compounded by the annual maximum of cloudiness in NW-region whereas in the greater Himalayas, cloudiness amount to three to four points only. Winter air throughout the region is dry; vapour pressure in the air-glacier interface is equal to 20-25 GPa. Thus, the annual amplitude of the air-glacier interface temperature from the Hindu Kush to the high Himalayas drops two fold from 20 to 10°C while vapour pressure in the same direction increases from 1-2 to 3-4 GPa.

The pressure of air circulation that determine the annual regime and distribution of cloudiness and air humidity result in•a proper change of the main factor responsible for glacier ablation viz, total solar eradiation The value of radiation on the glacier of the southern slope accounts for 90 percent of the maximum possible when the sky is cloudless and then declines in SE direction, reaching 60-65 percent on the glaciers of the Himalayas. At the same time, the absolute values of solar radiation influx in these areas are qual to 8.8 and 7.2 MJ/yr respectively. The daily snowfalls and abrupt rises of the surfacc albedo serve to further weaken the radiation factor of ablation. The frequent summer snowfalls in the Himalayas are due to monsoon. In the Hindu Kush, such snowfalls are much more seldom; for less than 25 percent of the summer season duration on the northern macroslope of the HIndu Kush they result from intrusions of cold air masses from the Atlantic and Arctic regions.

Fluctuations of glaciers are caused by changes in the climatic system. These fluctuations strangely affects the regional hydrology of glaciated mountains. The general shrinkage of mountain glaciers, following the end of the Little Ice Age (LIA), is a major reflection of the fact that rapid secular changes in the energy balance of the earth surface are taking place at a global scale. This glacial signal of atmospheric warming not only concerns the recent past but must be considered as one of the key parameters for early detection of man induced warming in the near future. Within the complex chain of processes linking climate and glaciers, glacier mass balances are most directly related to changes in atmospheric conditions. They are, therefore, of a special interest in view of the climatic trends which could exceed the band-width of variations within coming decades.

3

Importance of Himalayan Glaciers

The Himalayan snow and ice over Himalayan glaciers form cold natural water reservoirs which regulate the water supply throughout the year to our northern river systems. Water yield from a Himalayan catchment area is roughly double that of an equivalent catchment area from peninsular India, due to higher precipitation on the high mountains.

In any one year, glacier runoff may be greater or less than that year's precipitation because glacier runoff increases if precipitation is deficient (drought year) and decreases if the precipitation is greater than normal (flood year). Thus glaciers act as buffers and regulate the runoff from high mountains. Studies have shown that a glacier cover of 10 per cent of a basin area causes the stream flow variability to drop to half that of non-glacierized value (Meier and Roots, 1982). Basic requirements for understanding and proper management of mountain water resources requires knowledge of the amount and location of stored frozen water, the pattern of water release by melting, and how these patterns of release depend on short term weather and long term climatic change.

Glaciers are sensitive indicators of climate and its changes and have been used to resurrect past climates from the ice core studies. It is not possible to relate the glacier fluctuations to climatic variations directly. We can, however, measure glacier ice balances over large areas from aircraft or satellite, and we can learn something about climate history by observing the annual sequences in ice cores.

Briefly we can state the importance of Himalayan glaciers as follows :

- They exert a great cooling or refrigerating effect on the surroundings.
- They constantly interact with weather and climate both in temporal and spatial domains.
- They work as water supply reservoirs releasing more water under drought conditions of less precipitation and less under flood conditions caused by large precipitation.
- Due to their high altitude, their water can be advantageously used for production of cheap hydropower.
- The enormous snow and ice fields not only act as reservoirs for terrestrial and aquatic species but also for snow micro-organisms. The melting of snow and ice covers, the retention of liquid melt water and the release of soluble nutrients in the high altitude environment stimulate the growth of complex microbial communities influencing the bio-diversity of the region (IAHS,1994).

3.1 Cooling Effect

The vast high plateau of Tibet might have played a large role in reducing earth's temperature around 40 million years ago that triggered the Ice Age as pointed out earlier. The uplift of Himalayan Plateau cooled the whole globe including polar regions. Glacier activity on other mountainous regions increased and added to the worldwide level of erosion. The greater the erosion, the greater the cooling effect (Paterson, 1993).

The presence of snow and ice exerts a cooling effect on the surroundings affecting air circulations. As the moist wind rises over the mountains, it is cooled down and in presence of snow and ice, it is further cooled down and the moisture condenses. This explains the higher precipitation in glacierized areas. The increased precipitation, because of refrigeration effect of glaciers, may, therefore, be one of the dominant facts responsible for the extraction of the moisture and rendering the air mass fairly dry before it crosses the glacierized ridges. Snow and ice melt further enhanced cooling due to absorption of heat energy from the environment.

The redistribution of the cooled air mass from glacierized to other areas sets up wind patterns in the vicinity, modifying the local weather. The warmer air from adjoining areas also rises to the glacier regions and provides more heat for the generation of melt and, in turn, leads to generalized lowering of the temperature all around (Vohra, 1978).

3.2 High Altitude Effects

Some of the common effects of increasing altitude are decrease in air pressure, temperature, absolute humidity and density; increase in intensity of insolation (coming radiation), diurnal range of temperature, wind speed, and rate of evaporation; diminishing proportion of pollutants; increase in cloudiness and precipitation up to a certain level and increase in snow to liquid precipitation ratio.

Despite higher intensity of insolation, the capacity of atmosphere to retain heat is lower at high altitudes. The ground temperature is more than air temperature in day time which is reversed during night. The diurnal variation in temperature increases with altitude, which is the one of the main causes of weathering of the exposed rocks.

The increase in orographic precipitation ranges from less than 10 mm/100 m to over 300 mm/100 m under extremes of continentality to maritime environmental conditions. The precipitation follows the topographical conditions and it is observed that the first maxima of rainfall occurs near the lower Himalayas (~1200 m); the second near the middle range (2000-2400 m) and probably the third maxima occurs on the windward side of the Great Himalayan range near the permanent snowline (5000-5500 m), thereafter there is a substantial reduction in precipitation and cold arid to semi-arid conditions prevail (Bahadur, 1983).

The factors controlling glacierization of an area include the height of ridges, the orientation of slopes and the amount of precipitation in the area. In Himachal Pradesh, the minimum height of a ridge with a glacier is often about 5000 m while in eastern Nepal it is often 6000 m. South-facing sides of ridges have less ice cover due to strong insolation as compared to north-facing. It is observed that the glaciers descend to about 3700 m in eastern Kashmir (Sarbal glacier), 4000 m in Central Himalaya (Gangotri glacier) and 4500 m in eastern Nepal and Sikkim. The Khumbu glacier descending from Mt. Everest (8848 m), terminates at 4900 m while Kanchenjunga glacier terminates at 4200 m (Vohra, 1981).

3.3 Latitudinal Aspects

The solar and net radiation and temperature broadly decrease with increasing latitude. The latitude factor is apparent in the relative importance of seasonal and diurnal climatic rhythms. Seasonal changes of solar radiation, day length and temperature are small in low latitudes whereas the diurnal amplitude of temperature is relatively large (Barry, 1981). As we go to higher latitudes, the summer day becomes appreciably longer than the winter day and ground receives more insolation than at equator in summer months. Other latitudinal differences in mountain climates arise due to characteristics of the global atmospheric circulation. Tropical mountains are within the easterly trade wind regime, mid-latitude mountains are within the westerlies. The tropical easterlies decrease with height whereas the westerlies increase steadily. Associated with these differences in the global wind belts, precipitation systems are primary convective and small scale in low latitudes, but cyclonic and generally large-scale in middle and higher latitudes. As the Himalayas extend from 27°N to more than 35°N, the lower latitudes get affected by tropical and sub-tropical systems while the higher latitudes get affected by temperate and Arctic systems. The middle ranges have the mixture of both these situations.

Atmospheric circulations and the different latitudes have created Mediterranean maritime conditions in western Himalayas and monsoon conditions (both maritime and continental) in eastern and central Himalayas including Tibetan Plateau. Over a period of time, the environmental conditions have resulted in creating cold glacier region in north-western region and warm glaciers in north-eastern region of the country (Niang, 1990).

3.4 Glacier Inventory

We do not have the correct estimate of glacier area draining in our Himalayan rivers as their high altitude catchments lie further north in foreign territories. First generation inventory was made by GSI in 1983 showing that the country has 4795 glaciers having a glacial cover of 38,479 km². This has been updated showing 5243 glaciers covering 37,959 km² (Table 3.4.1) lie in our territory (Kaul, 1999). Probably 50,000 km² of glacier cover drain into our landmass making our rivers perennial and supporting a large population of men and animals.

Yet another effort was made by Survey of India (Issar, D.P., 2000) to prepare an inventory of major glaciers of the Indian Himalayas. This inventory is prepared statewise. 327 major valley have been identified. Out of these 60 glaciers are from J&K and Ladakh, 85 from H.P., 62 from Uttranchal and 20 from Sikkim Himalayas. Distribution of these glaciers are shown on 1 : 1 million scale.

A recent inventory of glaciers, glacier lakes published by ICIMOD of Bhutan (Mool, P.K. *et al.*, 2001) and Nepal (Mool, P.K. *et al.*, 2001) in cooperation with United Nations Environment Programme Regional Resource Centre — Asia and the Pacific (UNEP/RRC-AP) shows the followng salient information to be noted by the water resource manager for food control in India as these drain into our river systems (Table 3.4.2).

Table 3.4.1: Status of Glacier Inventory of the Indian Himalaya*

1. INDUS BASIN

Shyok, Nubra, Indus, Gilgit, Kishanganga, Jhelum, Chenab, Ravi, Beas, Satluj

	Basins	*Number of Glaciers*	*Glacierised Area (km²)*	*Ice Volume (km²)*
A. Detailed assessment (selected basins)	Jhelum Part of Satluj	133 224	94.18 420.60	3.30 22.99
B. Regional assessment	Remaining Basins	3398	33,382.00	
Total		3755	33,896.78 (Say 33,897)	26.29

2. GANGA/BRAHMAPUTRA BASIN

Alknanda, Bhagirathi, Kaliganga, Yamuna, Tista, Brahmaputra

	Basins	*Number of Glaciers*	*Glacierised Area (km²)*	*Ice Volume (km²)*
	Bhagirathi	238	755.43	67.02
	Tista	449	705.54	39.61
	Brahmaputra (Arunachal Himalaya)	161	223.37	9.96
A. Detailed assessment (selected basins)	Remaining Basins	640	2378.00	
B. Regional assessment		1488	4062.34 (Say 4062)	116.59

TOTAL

Total Number of Glaciers identified	:	5243
Total Glacierised Area (After integrating A&B)	:	37,959 km^2
Ice Volume (For Glaciers at A)	:	142.88 km^2

**Source* : Kaul, M.K., 1999.

Table 3.4.2: Summary of Glaciers and Glacial Lakes in Bhutan and Nepal*

Country	*Glaciers*			*Glacial Lakes*		
	Number	*Area (km^2)*	*Ice Reserves (km^2)*	*Number*	*Area (km^2)*	*Potentially Dangerous*
Bhutan	677	1316.71	127.25	2,674	106.776	24
Nepal	3252	5323.89	481.23	2,323	75.70	20

**Source* : Mool, P.K. *et al.*, 2001.

Preparation of glacier inventory of the Indian Himalayas is a difficult task as the glaciers occupy trans-national boundaries where aerial photographs are not available for various river catchments. The job can be done by seeking regional and international collaboration for the purpose so as to make realistic estimations of glaciers feeding the Indian Himalayan rivers. It is gralifying to note that ICIMOD in coordinating the inventory of glaciers, glacial lakes and the identification of potential glacial lake outfirst floods affected by global morning in the mountains of India, Pakistan, China/ Tibet Autonomous region under a project in cooperation with United Nations Environment Programme. The project is founded by the Asia-Pacific Network for Global Change Research, Japan. According to the present investigator, probably 50,000 km^2 of glacier area drains into our landmass making Indian Himalayan rivers perennial which support a large population of men and animals (Bahadur, J., 2002).

3.5 Glacier Fluctuations

The wide altitudinal range and the diverse orientations of the glaciers in the Himalayas and Trans-Himalayas make the region as excellent sites for study of glacier fluctuations.

On the basis of physical characteristics, the Himalayan glaciers are divided into two types — longitudinal and transverse. Longitudinal glaciers flow generally in the wider east-west trending valleys of the range. These glaciers are relatively long and have relatively gentle surface slopes compared to the transverse glaciers. The transverse glaciers with their shorter length and steeper surface slopes flow perpendicular to the main axis of the range.

Occasional and incidental observations made on the snout positions show that the glaciers of Himalayas and Karakoram mountains are under recession (Vohra, 1981). The retreat rates during the period of observation are given in Table 3.5.1:

Table 3.5.1: Retreat of Glaciers in the Himalayas and Karakoram*

Mountain Ranges	*Name of the glacier*	*Periods*	*Rate of retreat M/yr.*
Himalaya	Milam	1849-1957	12.5
	Pindari	1845-1966	23.5
	Shankulpa	1881-1957	6.8
	Poting	1906-1957	5.1
	Glacier No. 3 in Arwra Valley	1932-1956	8.3
	Gangotri	1935-1976	14.6
		1962-2000	35.3
	Zemu	1909-1965	7.9
		1975-1990	19.8
	Barashigri	1940-1963	44.3
	Sonapani	1906-1963	15.9
Karakoram	Minapin	1906-1929	21.8
	Biafo	1861-1922	0.0
	Kichik Kumdan	1946-1958	101.6
	Siachen	1929-1958	31.5
	Yengutsa	1892-1925	125.3

**Source*: Vohra (1981).

In 1979, Himalayan and Trans-Himalayan Glaciers fluctuations were studied since AD 1812. Local and regional synthesis of 112 of these fluctuation records are presented in this study (Mayewski and Jeschke, 1979). The regional synthesis deals with the fluctuations of glaciers in Kanchenjunga–Everest; Garhwal, Lahaul-Spiti; Kolahoi; Nanga-Parbat; Karakoram (North and South sides), Rakaposhi–Haramosh, Batura Mustagh, and Khunjerab Ghujerab. In a gross regional sense, Himalayan and Trans-Himalayan glaciers have been in a general state of retreat since AD 1850. Filtering of the fluctuation record with respect to glacier type and regional setting reveals that the period AD 1870 to 1940 was characterized by alterations in dominancy of retreat, advance and standstill regimes.

From the above glacier fluctuation, field evidence exists which indicate a lack of synchronicity in the commencement of the recession from Little Ice Age (ca AD 1550-1850) maximum. Some glaciers appear to have begun recession early in 19th century and still others appear not to have receded or even re-advanced. This behaviour is contrary to glaciers located in high latitude regions.

3.6 Comparison with Polar Glaciers

The three major cold environments under discussion are the Arctic, the Antarctic and the Himalayas (Bahadur, 1993). The Arctic is an ocean basin almost completely surrounded by continents. It is sometimes referred to as a "Mediterranean" in the continental hemisphere of the world. The Antarctic is a continent surrounded by ocean and is known as a "remote outpost" in the ocean hemisphere of the world. More simply the Arctic is called as a "hollow" and the Antarctic as a "hump". The Himalayas is the highest but the youngest mountain complex on the earth running in an arcuate shape for about 2500 km. It contains more than 530 peaks above 6000 m and contains about 50 per cent of all glaciers outside the polar environments. Following Table 3.6.1 gives a quick comparison of three great cold regions affecting climate on earth and its environment.

3.7 High Altitude Lakes in Himalayas

Both fresh and salt water lakes exist in high altitude regions in Himalayas. The lake water originates from snow and glacier icemelt contributions. The high altitude lakes occur at altitudes 3,500 m to 5,000 m having depth range from few metres to about 100 m. The areal extent of lakes vary from less than a hectare to several hundred hectares. The lakes are formed across the land of drainage by avalanche debris from

Table 3.6.1: Comparison between Himalayan and Polar Glaciers

Features	*The Himalayas*	*The Arctic*	*The Antarc*
Glacier area ($10^2 km^3$)	~95	134	50
Volume ($10^3 km^3$)	~1000	30	10
Average Elevation (m)	~6000	>100	~2400
Highest Elevation (m)	8848	3700	5140
Average Depth of ice (m)	~500	~1000	~2000
Velocity of ice flow (m/yr)	>>10	>1 to <1000	>1 to <10(
Average winter temp. (°C)	–36 to –1 (West to East)	–40	–60
Averages summer Temp.(°C)	41 to 23 (West to East)	5	–20
Lowest temp. recorded (°C)	–47	<-50	–89.2
Average windspeed (km/hr)	~100	<100	~100
Average precipitation (mm)	~500	25	189
Meltwater contribution (km^3/yr)	>1000	~1400	~2300

Source: Bahadur, J. (1992).

side slope or by the advance of a side glacier with its lateral moraines. The lakes formed by glaciers disappear totally during the dry season and in spring time and become refilled during summer months (June to August). The Himalayan lakes exist from cold deserts of NW Ladakh to wet and humid NE Imphal. The wide climatic variability introduces ecosystem variations that are unmatched in their extent and scope. Kashmir lakes are important for coldwater trout fish culture. The inflow of high silt load from glaciers is rendering the lake waters turbid and unfit for biological activity. Himalayan lakes can be divided into two classes — cirque lakes (tarns) which have developed above 3657 m and the structural erosional and rock dam lakes which have developed below 3657 m asl. Some prominent lakes are listed in Table 3.7.1.

Most of the Himalayan lakes suffer from wide environmental degradation. The changes in the lake environment are due to

(*i*) unregulated tourist flow;
(*ii*) flow of human, animal, agricultural and industrial wastes into lake waters;
(*iii*) conversion of lakes by hydro-electric dams;
(*iv*) siltation;

Table 3.7.1: Prominent High Altitude Himalayan Lakes

State	*Name of the Lake*	*Altitude (m) and type*
Jammu and Kashmir	Tarsai	3626 – Tarn
	Mansai	3640 – Tarn
	Gangabal	3570 – Tarn
	Sheshnag	3870 – Tarn
	Harmukh	3657 – Tarn
	Dal	1584 – Erosional Depression
	Wular	1580 – Erosional Depression
Himachal Pradesh	Tso Morari	4571 – Tarn
	Tso Kyagai	4782 – Tarn
	Tso Kar	4580 – Tarn
	Lam Dal	3992 – Tarn
	Mani Mahesh Dal	4169 – Tarn
Uttranchal	Kana Tal	4678 – Tarn
	Deosai	3413 – Structural Depression
	Khadru Danda	4847 – Tarn
	Shastri Tal	4847 – Tarn
	Lama Tal	4847 – Tarn
	Nag Kund	4602 – Tarn
	Naini Tal group	1950 – Structural Depression
	Devi Kund	4602 – Tarn
	Deo Tulea Tibba Tal	4574 – Tarn

**Source* : Ahmad, N. and Rais, S. (1998).

(*v*) invasion by exotic weeds; and
(*vi*) encroachment on lake catchment area.

There are only a few high altitude lakes where human impact is not significant. What is needed is restoration and preservation of these valuable lake ecosystems?

4

International Efforts for Study of Himalayan Glaciers

Due to enormity of Himalayan mountain system and its highest elevation on earth, it has attracted the attention of one and all and more so of the scientists of adventure for the study of the apex environment.

The first Chinese investigation (Pe/ we/, 1981) was undertaken in 1950 in connection with railway construction in NE-China (Manchuria). Permafrost problems were encountered in the highway construction, water supply and coal mining. University of Lanzhou introduced a permafrost course in 1963-64. Since 1978, the Lanzhou Institute was designated as the Institute of Glaciology and Cryopedology (now called as Institute of Glaciology and Geocryology) having research divisions on glaciology, cryopedology, mud-rock flow, surveying and mapping, material analysis and remote sensing telemetry.

Observations made for over two decades for about 500 mountain station (including 30 stations in glacier affected areas) in Tibet (Niang, 1990) gives the following salient data:

- average annual precipitation ranges from less than 20 mm to more than 5000 mm.
- variation coefficient for the annual precipitation is 20 per cent in eastern slopes, 20-30 per cent in northern slopes and 30 to 70 per cent in southern slopes.

- high mountain belts have stable precipitation with maximum at 5250 m, i.e. near the equilibrium line on the glacier.
- Annual evaporation in glacier region varies from 70 to 120 mm/yr.
- Changes in snowline are higher in eastern than in the western region.
- Negative temperatures are observed over continental glaciers for over 6 months as compared to 3 to 4 months on maritime glaciers.
- Glaciers in SE-Tibet ablate 3 to 6 m/yr.
- Solar radiation is the main source of heat next to induced heat (< 5%).
- Hydrological year for continental glaciers is from September to October and for maritime glaciers, it is from January to December.
- Runoff modulus for glaciers :
  25.5 to 61.0 l/(s km^2) for continental glaciers
  85 to 195 l/(s km^2) for maritime glaciers.
- Annual glacier melt contributions are about 50 per cent in the west which reduces to about 10% in north-east.
- Rivers with large glaciers have less danger from droughts, i.e. deficit or excessive rain, causing floods.

Chinese have been actively collaborating with Japanese and Americans since 1981 onwards to investigate the complexity of the high altitude Himalayan environment.

4.1 Joint Chinese and Japanese Studies

Joint Chinese and Japanese glaciological investigations on Qingzang Plateau, 1981, were undertaken to investigate existing conditions for glaciological phenomena such as glaciers, and also to investigate the role of other forms of snow and ice, that is snow cover and permafrost in the water cycle and energy budget in this region. Characteristic features were occurrence of abundant snowfall, partly due to large scale disturbance and due to local convection (precipitation from cumulonimbus clouds). The $\delta^{18}O$ - value of precipitation shows a possibility of precipitation of local origin water vapour. Formation of snow cover resulted in cooling of the atmosphere and retardation of deepening of permafrost active layer. Abundant evaporation was observed from snow cover and glacier surface. An automatic meteorological observation system is maintained in this region to collect long term data.

Preliminary results from China-Japan Glaciological Expedition in Qing-Zang (Tibet), 1989 are :

- From shallow ice-core records and meteorological data, climatic warming trend in the Tibet Plateau is evident resulting in intense glacial retreating in SE Tibet and also resulted in obvious decrease in glacial velocity on the basis of reconstruction of glacial ogive.
- The glacial fluctuation trend in the East Kunlun and the Tanggula area is different from that in SE Tibet. Many glaciers show evidence of glacial advance.
- $\delta^{18}O$ analysis snow samples from precipitation process indicate that there is a big difference in $\delta^{18}O$ between maritime air mass and continental air mass. This can be used to distinguish the origin of precipitation.
- Precipitation in the region are mainly high temperature connected with high precipitation and low temperature with low precipitation.
- The concentration of chemical composition (K^+, Na^+, Ca^{2+}, Mg^{2+}, SO_4^{2-}, NO_3^- and Cl^-) show marked oscillatory increase downward from the top in the ice core, suggesting that the concentrations of chemical composition in Tanggula ice core have a trend of oscillatory decrease from past to the present which is contrary to that observed from ice cores in polar regions when the trend is increased from past to the present.
- Since the start of the 1940, the alkalinity degree has gradually decreased and acidity degree has gradually increased. This has contributed to the decrease in

precipitation process connected with the alkaline dust. During 505 years, the conductivity gradually decreased in Tanggula mountain ice core. Air temperture was not the major factor that affected the changes of conductivity. The strong sources of dust and purification of air by precipitation were the major factor leading to changes in alkalinity.

- In dry and cold climatic conditions of polar and sub-polar region, the ice temperature of ice cap glacier at 10m depth is almost equal to annual mean air temperature.
- The influence of snow cover primarily has effect in increasing the surface albedo and maintaining the surface temperature at 0°C or below. This means cooling of air layer near the surface and resultantly, it will introduce various influences, e.g.:
  (*a*) Low ground surface temperature
  (*b*) Low wind speed
  (*c*) Global Solar Radiation (G. S. R.) shows inverse relation with cloud coverage (cc) for snow-free period. When snow cover existed, the GSR show high value although cc is high, meaning different tendency.
- Run off of melting snow in spring, of the streams in headwater of Yangtze river amounts for 8 per cent of annual volume of run-off.
- Glacial changes of Tibet since last glaciation on the basis of glacial remains — Radio carbon datings and former snowline altitude reconstruction taking examples of Kunlun Pass, Tanggula Pass and Bomi area (SE Tibet), there are at least several obviously glacial advancing stages since last glaciation, e.g. 22000-16000 a BP (Last Glaciation); 12000a BP (late glacial); 3000a B.P., 1900-1500a B.P., 1000a B.P., 1000a B.P. (Neoglaciation); 600a B.P., and 200a B.P. (Little Ice Age). The evolution of glaciers had the characteristics of glacial change extent in the margin of the Plateau being more than that in the inland of

the Plateau. During the last glaciation, the glacial scale in the margin of the Plateau was 4-7 times than that of existing glaciers, and snowline altitude descended about 700-1000 m; whereas the glaciers in the inland of the plateau was 1-3 times than that of existing glaciers and the snowline altitude descended about 100-500 m. The scale of glaciers are gradually reducing and snowline altitude are rising from the last glactiation to the present.

- Fluctuations of Zepu glacier in the Eastern Nyainguentanglha mountains during late Holocene Epoch using geo-morphological research and radio-carbon dating supports the following conclusions:
  (*a*) During the past 600a B.P., the Zepu glacier has advanced about 2 km from the present terminal position.
  (*b*) Neo-glacial advances of the Zepu glacier occurred prior to 1000a BP and the maximum advance attained down to 5 km from the present position.
  (*c*) Deforestation by the human activity might have taken place between 700 and 2000 B.P.
  (*d*) Preceding the Neo-glacial period, some large movements happened and thick debris covered the ablation area of the glacier.
  (*e*) In general, the late Holocene glacier fluctuations of Zepu glacier is in agreement with the standard fluctuation curve described by Scientific Expedition to the Qinghai-Xizang, Plateau, 1986.

4.2 Joint American and Chinese Studies

Under joint American and Chinese studies glaciochemical investigations were conducted on snow pits and ice cores from polar ice caps, viz. Dunde and Guliya from 1981 onwards.

Modern eolian dust deposition in Central Asia through the analyses of concentrations and size distribution of insoluble particles (over the size range 1-22 µm diameter) in snow and ice samples collected from eight glaciers in the mountains of Central Asia (Wake and Mayewski, 1994).

The sample locations cover a wide geographic area and fall within from distinct landscape (climatic zone) — southern slopes of the eastern Himalayas; the Karakoram; south-eastern Tibetan Plateau, northern-western Tibetan Plateau. Snow samples from each of these landscapes displays distinct dust deposition characteristics which are dependent upon the physiographic location of the sampling site w.r.t. source regions of dust and atmosphere transport pathways (Fig. 4.3.1). The salient results of these studies are as follows :

- The southern slopes of the eastern Himalayas show very low annual dust deposition. The region is relatively free from the influence of Asian dust.
- The Karakoram appears to be a major suite for particles less than 2 µm in diameter which have been transported long distances with the westerly jet stream.
- Glaciers in the south-eastern Tibetan Plateau record the influx of dust generated in the arid and semi-arid regions to the west.
- Glaciers on the northern and western margin of Tibetan Plateau lie adjacent to vast arid and semi-arid regions of western China and provide dust deposition records mainly of local significance.

Since 1987, ice-cores have been drilled from Dunde and Guliya ice caps on Tibetan Plateau. Dunde ice-cap is located in NE Tibet with average precipitation of about 1400 mm/yr and the ice-core was taken at 5325 m while Guliya ice-cap is located in NW Tibet with average precipitation of about 180 mm/yr and the core was taken at 6200 m (Lin *et al.*, 1995). Temperature profiles of both the ice caps shows that the ice caps are polar throughout the depth with negative temperatures up to the bottom.

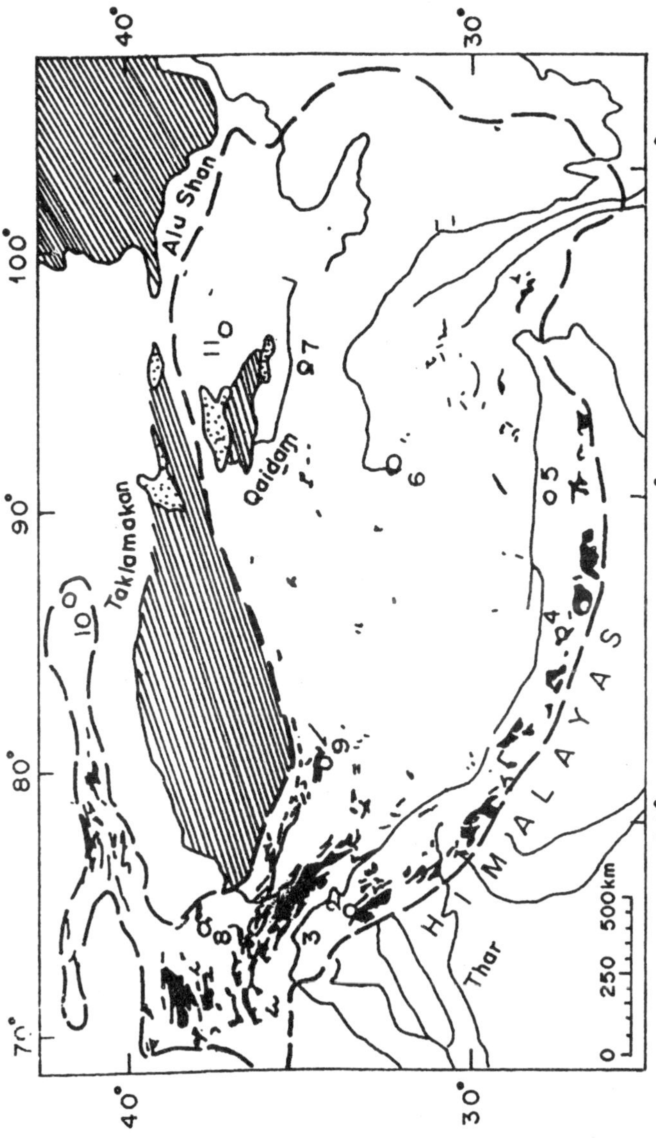

Fig. 4.3.1: Location Map of Glaciochemical Collection Sites in the Mountains of Central Asia

Oxygen isotope ($\delta^{18}O$) records for last 1000 years from both these cores are compiled and compared. Using surface temperature (Ts) observations since mid-1960 from meteorological stations on the plateau and $\delta^{18}O$ measured on precipitation collected contemporaneously, the empirical relation; $\delta^{18}O = 0.6Ts - 12$ is established. $\delta^{18}O$ appears to serve as a reasonable proof for regional surface temperature and a reasonable basis for reconstructing 1000 years proxy temperature records from Dunde and Guliya.

The reconstructed temperature histories for Dunde and Guliya show some centennial scale similarities, but reveal quite different histories for higher frequency variability over the last millennium. The ice-core $\delta^{18}O$ histories from Dunde and Guliya are compared with a tree-ring index from western China and dust fall record from eastern China, but shows no consistent relationship.

The most prominent similarity between the reconstructed temperature history for Dunde and Guliya is the marked warming of the last few decades from 1000a perspective, provided by these ice-core records; warming over Guliya is more recent (since 1985) and not unprecedented. The recent warming over Tibetan Plateau is evident in the limited meteorological records.

4.3 International Karakoram Project (IKP)

An International Karakoram Project (IKP) was launched in 1980 by Royal Geographical Society as a collaborative research project undertaken by the Chinese, Pakistan, Swiss and British Scientists to demonstrate the immense research potential of the area. Over 600 pages report was presented (K.J. Miller, 1984) and published by Cambridge University Press, Cambridge. The report represents only one beginning of an integrated research to unravel the natural mysteries to solve the human problems associated with the most chaotic landscape on the surface of our earth. Several groups reported on geology, glaciology, survey earthquakes, housing and natural hazards including geomorphology. The basic philosophy that

1. Perennially snow covered Bhagirathi II mountain pea (6454 m) (*Photo:* IMF, Delhi)

2. Snow Profiling for density measurements for evaluation of its water content.(*Photo:* Author)

3. Satellite photograph of snow and ice covered region of Himachal Pradesh with Govindsagar lake (Bhakra reservoir) on river Sutlej. (*Photo:* NARSA, Hyderabad)

4. Landsat imagery of snow covered region in Nepal Himalayas feeding Ganga river system.

5. Space Imagery of Everest region with Khumbu glacier system.

6. Alpine tents and snow conditions in the beginning of October, 1971 at Gaumukh camp site during an Interdepartmental Gangori Study Expedition (The author is in dark blue jacket, standing second from left).

7. The members of the expedition are clearing the snow to retrieve kitchen tent which was buried under the heavy snow which fell continuously for three days in early October, 1971. (*Photo:* Author)

8. Meteorologist-member of the 1971-glacier expedition was concerned to forecast the weather conditions for clearing the overcast sky for the safe return of expedition team to Uttarkashi.

9. A view of ice-tunnel with emanating Bhagirathi river from Gaumukh. (*Photo:* Author)

10. A panoramic view of perpetually snow covered Bhagirathi peaks (I-6856m, II-6454m and III-6512m) with Gaumukh @ the snout of Gangotri glacier.

11. Glacier ice enroute Kamet peak (7756 m). (*Photo:* IMF, Delhi)

12. Debris covered Dokriani glacier with its snout in Gangotri region.

13. Closer view of snout of Dokriani glacier. The ice is heavily fractured and broken and is decaying fast due to global warming. (*Photo:* Glacier Research Group, J.N.U., New Delhi)

14. Glacier lake formed by damming of drift and other rock deposits. (*Photo:* Author)

15. Heavy rock and soil erosion by strong freeze@thaw cycle. (*Photo:* Author)

16. Rushing and turbulent snow and glacier melt fed m o u n t a i n s t r e a m is being crossed by ropeway at Nandkarak. (*Photo:* Glacier Research Group, J.N.U., New Delhi

guided the sponsor of the IKP, namely that the Karakoram mountains of Pakistan represent a laboratory for world science and that all who go to this area should acknowledge that the great forces that distort the surface of our earth can be equalled by man's energy and ability to work together to the benefit of all (Miller, 1984). The major successes of the project are listed below:

The Major Successes of the International Karakoram Project

1. Determination of the correct survey stations on Haçhindar, Buriharar and Shunuk.
2. Discovery of the 1913 Atabad Survey Station.
3. Positioning of key locations by satellite receiver equipment to enable rapid future advances by the Survey of Pakistan (SoP).
4. Positioning of most of the major seismic stations by satellite Doppler equipment.
5. First ever 3-Dimensional recording of the shape, size and position of Karakoram Glaciers (Hispar and Ghulkin), a first step towards documenting a natural resource and a hazard.
6. The first multi and inter-disciplinary study of the vulnerability of low-cost housing and settlements to natural hazards.
7. Development of a new methodology to assess natural hazards.
8. A detailed critical assessment of the health and medical care provided for the people of the Karakoram.
9. The exposure of priority needs in villages especially the need for prevention before a catastrophe, rather than aid after the event.
10. Establishment of the clear need for education on:
    (i) Clean water supplies,
    (ii) Eradication of poor building styles,
    (iii) Integration of social/medical services,

(iv) Deforestation control.

11. The need for operational research procedures to the maintenance of the Karakoram Highway.
12. The establishment of the need to have an interdisciplinary approach by geomorphologists, architects and engineers to the design and safe location of new buildings.
13. A report on the structural repair work urgently required to preserve the most historic building in the Karakoram, Baltit Fort.
14. The first detailed seismological study ever carried out in the Karakoram.
15. The most thorough series of geomorphological investigations ever performed in the Karakoram that will assist a more precise documentation and hence a wider appreciation of the natural hazards faced by the indigenous population.
16. A continuation of geological surveys in what is, and will long remain, a relatively unexplored area in geological terms.
17. The first detailed study of the glacial sedimentology of the Upper Hunza region.
18. The first thermoluminescence datings of Pleistocene and recent sediments in the Karakoram.
19. The first coring of quaternary lake sediments in the Upper Hunza Valley.
20. An extension of known quaternary glacial stratigraphy of the Hunza Valley.
21. Detailed mapping of glacier snout positions and fluctuations in the Hunza area.
22. Provision of detailed water chemistry data for the Hunza river and its tributaries.
23. Measurement of high altitude rock temperatures.

Collaborative Ventures

24. Close collaboration with several government agencies in Pakistan, namely the Water and Power

Development Authority, The Survey of Pakistan, The Geological Survey of Pakistan, The Frontier Works Organisation of the Pakistan Arm and the Kohistan Development Board.

25. Close collaboration with, and lectures to, Pakistan Universities at Karachi, Quetta, Punjab, Engineering Technology of Lahore, Peshawar and Quaid-I-Azam Islamabad.
26. Lectures to the Engineering Design Staff of the Pakistan Atomic Energy Commission.
27. Lectures to the Metropolitan Planning Board in Karachi.
28. Successful collaboration with several institutions in China including the Institute of Glaciology and Cryopedology, Lanzhou, Lanzhou University, Beijing University, Four other Institutes of Academia Sinica.
29. Continuing and long-term co-operation with all Pakistan and Chinese collaborators.
30. A continuation of assistance by sponsors, particularly the British Council, to those associated with the project, FG, long-term liaison with overseas universities.

4.4 Snow and Ice Hydrology Project (SIHP)

Under Snow and Ice Hydrology Project (SIHP) during 1981-85 between Pakistan's Water and Power Development Authority (WAPDA) the International Development Research Centre (IDRC) in Ottawa and various Canadian universities, glacio-hydrological aspects of upper Indus river basin were explored (Hewitt and Young, 1992). In Pakistan, project coordination was undertaken by Hydrology Research Directorate (HRD) of WAPDA in Lahore. In Canada, the project was coordinated within the Cold Region Research Centre at Wilfred Laurier University (WLU) Waterloo, Ontario. Scientists from the universities of Waterloo, Ottawa, British Columbia and the Inland Waters/Land Directorate of Environment, Canada also contributed. The universities of

Manchester (UK) and New Hampshire (USA) provided additional expertise. The main objectives of the project were to :

(i) Initiate research into glacio-hydrologic aspects of the upper Indus basin relevant to water resources development and forecasting.
(ii) Introduce and test technical methods and models appropriate to the environment of the Himalayan Indus.
(iii) Define the terms of an ongoing monitoring and forecasting system for snow and ice regime basins.
(iv) Train a core of Pakistan scientists/engineers to continue this work. The project recommended trials and establishment of high altitude (2500-6000 asl) monitoring stations and instrumentation at suitable locations.

4.5 Ev – K^2 – CNR Scientific Project

Italian National Research Council (CNR) launched the Ev - K^2 - CNR Scientific Project in 1990 for multi-disciplinary researches in earth, human, environmental, medical, biological and human sciences and technological research. To host the scientific laboratories, a Pyramid of glass and aluminium was established on southern slopes in Nepal in cooperation with the Royal Nepalese Government. It has been installed at 5050 m altitude in the Sagarmatha National Park, close to a glacial lake. In 1991, a broad band seismic station was established. Italian National Research Council, (Ev – K^2 – CNR Scientific Project, 1990).

4.6 Regional Programme on Mountain Hydrology

A Regional Programme on Mountain Hydrology was jointly considered by UNESCO, WMO and ICIMOD by a working group in 1992 with no Indian participation. The main object of the project is to promote and enhance indigenous capability for sustainable development of water resources in the Hindu-kush Himalayan Region (Chalise, 1993).

The other important objective is to provide exchange of data, information and knowledge between the participating institutions to help develop appropriate hydrological models for diverse watersheds to provide protection to human settlements and infrastructures in the downstream regions.

4.7 Himalayan Interdisciplinary Paleoclimate Project (HIPP), 1995

The project workshop was sponsored by the US National Science Foundation (USNSF) and IGBP (International Geosphere and Biosphere Programme)–PAGES (Past Global changes) to synthesise and stimulate an inter-disciplinary investigation of Paleoclimate in high lands of Central Asia. Interest on several lines of environmental modelling where discussed as avenues for future collaboration among scientists working in the Himalayan region. Some of highlights are as follows:

1. Understanding the role of Himalayan/Tibetan uplift in modulating broad-scale and local climate change in the region. As with many other potential model applications in this topographically- complex region, the need was stressed for very high resolution (e.g. nested limited area) climate models — the development of such models should receive high priority.
2. Understanding how well state-of-the-art climate models simulate the dynamics of monsoon variability across inter-annual to millennial time scales. One opportunity for interaction is with PAGES Paleoclimate Modelling Inter-comparison Project, which is coordinating the simulation of specific time periods by many different modelling groups. Initially, the exact same initial/boundary conditions are being used by over 12 modelling groups to simulate 6 and 18 ka. It was concluded at the workshop that a high

priority should be placed on understanding the patterns of the climate over the Himalayan region at these two times, and in particular, that scientists working in this region should coordinate with others working outside this region to understand how well our predicted models simulate observed climate changes of the past. It was also recognised that "forward" models will be needed to transform climate model simulations into a form comparable with proxy data. For example, climate model output can be used with vegetation models to infer past vegetation change.

3. Understanding the role of different climates forcing mechanisms (e.g. Milankovitch, trace-gas, ice volume) on monsoonal climate change over the last glacial cycle. Equilibrium climates simulations will be made for periods of specific forcing, and then compare to array of proxy data for the same time slices.
4. Understanding the relative roles of various natural and anthropogenic climate forcing mechanisms over the last several centuries. Models simulations are being carried out using reconstructed solar, volcanic, trace gas and aerosol forcing over the past 500 years. The patterns of simulated change will than be compared with observed paleo-environmental time series on global basis. A high priority was placed on generating annually-dated time series in the Himalayan region that can be compared to the model simulations to unravel the possible causes of observed inter-annual to century-scale climate variability.
5. Given the importance of hydrological change in the Himalayan region, and the widespread use within (e.g. in ice cores and sediments) of stable isotopic paleo-climate proxies, the need was emphasised for the development of appropriate hydrological models, and the use of isotopic tracers in climate simulations to understand the sources and pathways of moisture into the Himalayan region in the past.

4.8 Global Change Impacts on Mountain Hydrology and Ecology (IGBP-BAHC/GCTE), 1996

A workshop was organised in April 1996 at Kathmandu (Becker, A. and Bugmann, H., 1996) under IGBP (International Geosphere Biosphere Programme) the objectives of the workshop were :

1. to define the key global change-related scientific questions in hydrology and ecology that can be addressed with sharp environmental gradients based on topography.
2. to develop a scientific "Mountain Work-plan" for coordinated BAHC (Biological Aspects of the Hydrological Cycle)/GCTE (Global Change and Terrestrial Ecosystems) research in mountain regions with particular emphasis on (a) assessment and modelling of the interactive influence of topography and surface heterogeneity on the spatial pattern of soil moisture, evaporation, run-off generation and erosion; (b) collaborative ecological, micro-meteorological and hydrological process studies along altitudinal gradients coupled, where feasible, with integrated hydrological studies; (c) measurements, manipulative experiments and modelling studies of ecosystems dynamism along altitudinal gradients; (d) requirements for measurement techniques in harsh mountain environments.

The workshop identified three main activities and each activity was subdivided into three research tasks listed below:

Activity 1: Study Hydrological and Ecological Processes and their interactions along altitudinal gradient as a function of land surface characteristics.

1.1 Monitoring and manipulative experiments along altitudinal gradients for understanding hydrological and ecological process and their interaction.

1.2 Assessment of the spatial and temporal variation of hydrological and ecological processes and their interaction in mountain regions as a function of climate and land surface properties.
1.3 Modelling the impacts of climatic and land use changes on hydrological and ecological processes in mountain region.

Activity 2 **:** Develop measurement protocols and models to use ecological and hydrological characteristics of mountain regions as indicators of global change.

2.1 Identification of suitable hydrological and ecological indicators of global change impacts along altitudinal gradients.
2.2 Development of methodology for the coordinated use of altitude-related indicators to identify global change and its impacts.
2.3 Establishment of global monitoring programme.

Activity 3 **:** Develop concepts for sustainable development based on relations between socio-economic conditions and ecological/hydrological characterises at different altitudes in mountain regions.

3.1 Quantification of the altitude-specific relationship between socio-economic characteristics in particular land use and hydrological and ecological processes.
3.2 Develop altitude-specific management practices for mountain ecosystems which ensure sustainable development and environmental protection in the mountain regions as for down-stream communities.
3.3 Assessment of interactions between ecosystems structure, land use practices and extreme events (floods, droughts, erosion, landslides etc.) as a basis for development of protection measures and mitigation strategies.

5

Perennial Himalayan Rivers

The perennial rivers draining the Himalayan mountains sustain a huge population of men and animals in the northern parts of the Indian subcontinent including mountains and Indo-Gangetic plains (Fig 5.1). However, they cause floods in the raining season, year after year affecting human life and property. The people have accepted their recurrence are rather welcome it due to great beneficial effects of rejuvenating the environment by scavenging it and adding fertile soils for increased crops productivity.

5.1 Major River Systems

There are three major river systems, viz. Indus, Ganga and Brahmaputra.

5.1.1 Indus River System

The Indus river originates north of the Great Himalayas on the Tibetan Plateau, (Fig. 5.1.1) near Lake Mansarovar (~5182 m) and Mt. Kailash. (Negi, 1998). The river flows westward, south of Karakoram Range and north of the Great Himalayas, to Mt. Nanga Parbat, where it turns sharply to the south, flowing through Pakistan into the Arabian Sea near Karachi after travelling a distance of about 2880 km. The length of the river in India is about 1114 km. The principal tributaries of the river are the Sutlej, the Beas, the Ravi, the Chenab and Jhelum.

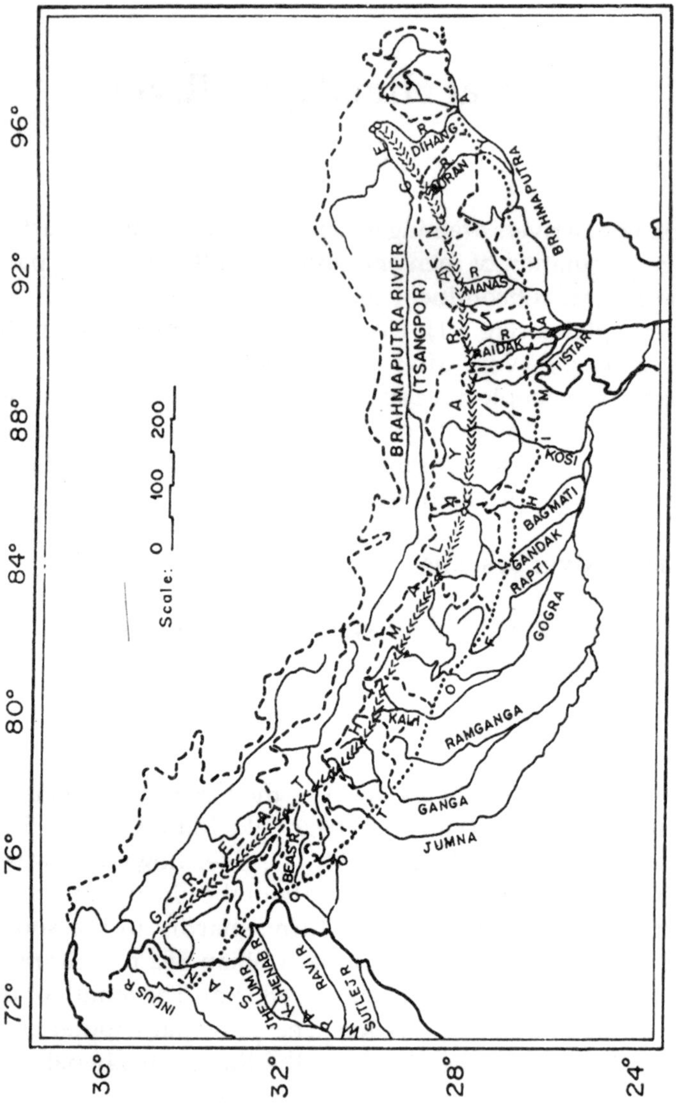

Fig. 5.1: Perennial Himalayan River Systems.

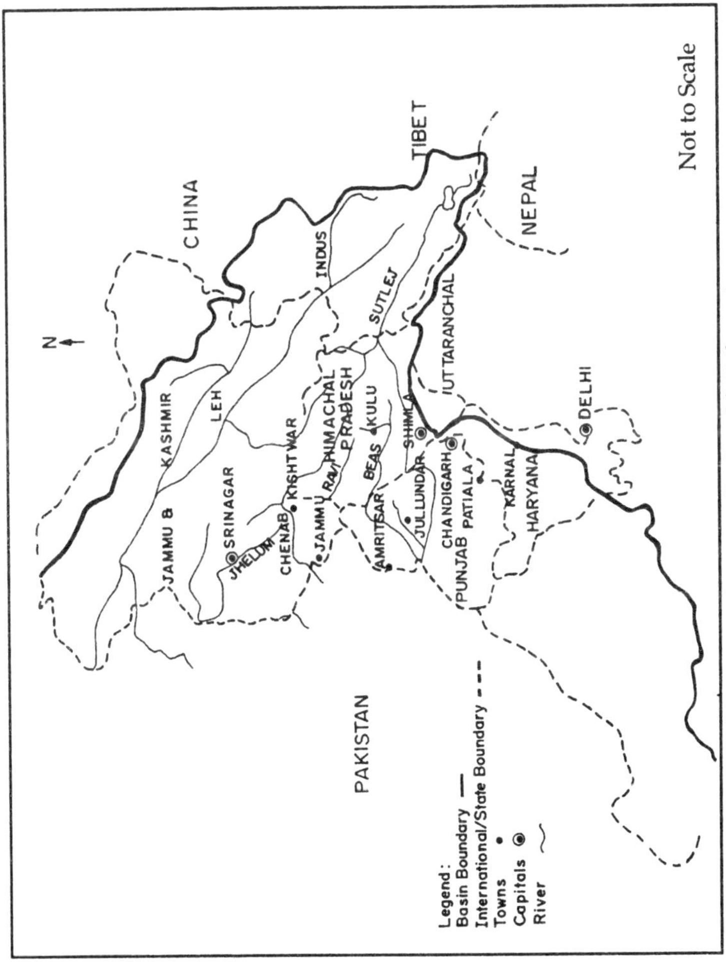

Fig. 5.1.1: Indus River Basin with Tributaries.

Sutlej

The river Sutlej rises near the lake Mansarovar in Tibet. It flows for a considerable distance before entering Indian territory near Shipki La. Thereafter it drains the trans-Himalayan zone of Spiti, its major tributary. The river Sutlej has cut across the great Himalayan range through a deep gorge. Just upstream of this gorge, it is joined by the river Baspa which drains the eastern part of Himachal Pradesh. After crossing the great Himalayan range, the river Sutlej flows in a more or less south-west direction before emerging into the plains near Bhakra.

Beas

The river Beas is another major tributary of the Indus river. It rises on the southern slopes of the Pir Panjal range just below the Rohtang (~4000 m). It drains Manali and Kulu valleys cutting Dhauladhar range at Larji. The river gradient from Rohtang massif to Larji is very steep. The main tributaries of the river Beas before it cuts across the Dhauladhar range are:

1. The Parbati which rises in the snowy wastes from Manikaran. It joins the river Beas near Shamshi in the Kulu valley.
2. The river Harla joins the Beas near Bhuntar.
3. The river Sainj which rises in the snows of an off-shoot of the Pir Panjal range that marks the watershed of the Beas and Sutlej rivers. It joins the river Beas near Larji.
4. The river Tirthan which rises in the snows of an off-shoot of Pir Panjal range and joins the Beas near Larji.

The gradient becomes gentler south of Larji. Thereafter the river Beas drains past the town of Mandi and takes a westerly course across the southern part of the Kangra valley. In Kangra valley, tributaries Uhl, Suketi, Luni, Awa, Banganga,

Gaj and Chak join the river Beas. The river Beas enters the plains of Punjab near Pathankot after cutting across Siwalik range.

Ravi

The Ravi is another important tributary of river Indus. It rises in a tract between the Dhauladhar range in the south and the Pir Panjal in the north. It originates in Bara Bangahai as a joint stream formed by the Bhadal and Tant Gari fed by glaciers. The river Ravi flows in more or less westerly direction before it cuts across the Dhauladhar range to enter the plains of Punjab. Its main northern bank tributaries are the snow-fed Siul and Baira streams.

Chenab

The river Chenab is another major tributary of the river Indus. It drains parts of north-western Himachal and western Himachal Pradesh and the eastern parts of Jammu and Kashmir. It is made up of the rivers Chandra and Bhaga, both of which originate on the north-western slopes of great Himalayan range. Both the rivers flow through sparsely inhabited region. The river flows through the famous Pangi tract of Himachal Pradesh. Its main northern bank tributaries are Saichu and Miyar streams which originate in the snowy wastes of the southern slopes of the great Himalayan range. The river cuts through the Pir Panjal range and crosses over from the north flank to southern part. It flows for a considerable distance along the base of the Pir Panjal range in Kashmir, before entering the plains where it joins the river Indus.

Jhelum

The river Jhelum is yet another major tributary of the river Indus. It originates in the northern slopes of the Pir Panjal range which girdles the valley of Kashmir. It flows past the centre of the valley before cutting through the Pir Panjal range in a narrow gorge from Baramula to Muzaffarabad where it turns sharply towards south making a hairpin bend. Its major tributaries are river Liddar and river Kishenganga. The river Liddar, originating in the southern slopes of the great Himalayan range which hems the Kashmir valley. It joins the river Jhelum near Srinagar.

The river Kishenganga also originates on the southern slopes of the great Himalayan range. It drains the NW part of the Kashmir valley. It cuts across the Pir Panjal range and joins the river Jhelum further downstream. A number of small tributaries draining the Kashmir valley also contribute their water to the river Jhelum.

5.1.2 Ganga River System

Ganga river basin parallels the long axis of the Himalayan Range from the headwaters of Himachal Pradesh, right from the eastern slopes of the Shimla ridge to eastern Nepal. The course of the river has not been determined by down-cutting the erosion, throughout much of its length, but rather by the down-warping of the crust in the "subduction zone" to the south of the tectonic uplift of the Himalayas.

The Ganga originates as Bhagirathi from Gangotri glacier (4000-5600 m). The Bhagirathi is joined by the Alaknanda at Deoparyag and the combined stream is known as Ganga. It debouches into plains at Rishikesh. It is joined by a large number of tributaries on both the banks in the course of its total run of 2525 km. before its outfall into the Bay of Bengal. Its major tributaries are Yamuna, Bhagirathi, Alaknanda, Kali, Ramganga, Ghagra, Gandhak and Kosi (Fig. 5.1.2.1).

Yamuna

The river Yamuna is the largest tributary of the river Ganga. It drains the western-most part of the Ganga catchment. It rises from Yamunotri glacier in Uttarkashi district lying on the SW-slope of the Bandrapunch peak. It flows for a considerable distance along a SW direction up to the base of the Mussorie

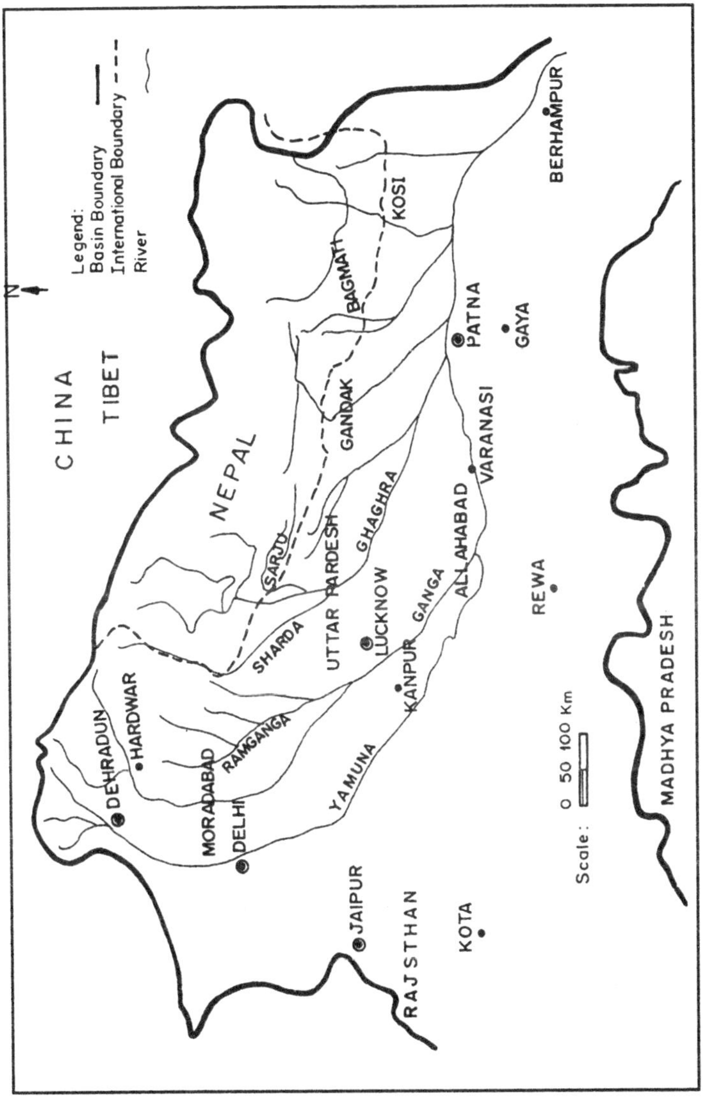

Fig. 5.1.2.1: Ganga River Basin with Tributaries.

ridge after which it turns towards west. The following are the main tributaries of the river Yamuna.

1. The river Tons rises in the snows beyond the valley of Har-ki-dun. It is made of two main rivers — the Rupin and the Supin. The river Tons then drains in a SE direction before joining with the river Yamuna at Naugaon.
2. The river Aglar drains the northern slopes of the Mussorie ridge.
3. The river Giri drains the SE corner of Himachal Pradesh. It flows past the base of Shimla ridge which marks the water-divide of the Indus and Ganga catchments. The river Giri then flows in a SE direction, through Sirmur and past Renuka before joining the river Yamuna near Paonta.
4. The river Bata originates in the lower slopes of the Nahan ridge. It flows through the Paonta valley in an easterly direction before joining the river Yamuna.

Bhagirathi

The river Bhagirathi rises from the south of the Gangotri glacier at the base of the Chowkhamba group of peaks. Thereafter, it flows in a SW, S and E direction past the towns of Uttarkashi. and Tehri to merge with the river Alaknanda at Deoparyag to be called Ganga. The river flows through a narrow gorge before emerging into the plains after cutting across the Shivalik hills near Hardwar. The river Bhilangana is the largest tributary of the river Bhagirathi. It rises from Kathling glacier (3800-6500 m) to the NE of Tehri and drains through a narrow river valley to join the river Bhagirathi near Tehri.

Alaknanda

The Alaknanda river rises in the glaciers to the north of the temple town of Badrinath. It traverses through the towns of Vishnuparyag, Nandparyag, Karanparyag, Rudraparyag and

Srinagar. The main tributaries of the river Alaknanda are Mandakini, Pindar, Nandakini and Dhauliganga.

1. The river Mandakini rises from the snowy wastes of the Bhagirathi Kharak and Satopanth glaciers to the east of Kedarnath. It is the biggest tributary of the Alaknanda river.
2. The river Pindar rises from the Pindari glacier on the western slopes of the Alaknanda-Kali water divide. It flows in a more or less westerly direction before joining the Alaknanda river at Karanparyag.
3. The river Nandakini is another important tributary of the Alaknanda. It flows along a SW direction before going into the main river at Nandprayag.
4. The river Dhauliganga rises from the snowy wastes on the southern face of the great Himalayan range. The river Pushpawati, draining the valley of flowers, is one of its main tributaries.

Kali

The river Kali forms the border between India and Nepal. It flows in a more or less SSW direction. There are two headwaters of this river: the Kalapani, the eastern headwaters which is a collection of spring and the Kuthi Yankti, between headwaters from snowfields of the Himadri on the southern slopes of the great or main Himalayan range. Both these headwaters merge to form the mainstream of the river Kali. Its main tributaries are Goriganga, Sarju, Ladhiya, and Lohawati. A number of tributaries join the river Kali from eastern flank from Nepal. The river Kali enters the plains at Baramdeo and further downstream, it is known as river Sarda. It flows into the river Ghagra in the plains of Bihar.

Ramganga

The river Ramganga drains parts of SW Kumaon. It is formed by a number of streams which join the main stream that drains a V-shapes valley, past the broad-terraced fields of Ganai and through the dense foothill forests of the Corbett National Park to enter the plains downstream of Dhikala.

Ghagra

The Ghagra river system drains western Nepal. The river Karnali is the largest river draining into Ghagra. The Karnali receives water from a large number of snowfed spring-fed tributaries. The river flows a considerable distance along a southerly direction before it turns towards SW and once again towards SE to join Ghagra. Seti and Bheri rivers are other tributaries. The river Ghagra or the Girwa and the Kauriala as it is known in Nepal, pierces the lower hills near Gaindakanda to enter the plains of Bihar.

Gandak

The Gandak system drains central Nepal. The main river of Gandak river is the Kali or the Krishna Gandak. It rises in the trans-Himalayan beyond Manang Bhot. The river has cut across the main or great Himalayan range through a spectacular gorge between the Dhaulagiri and the Annapurna massifs. The other important rivers of the Gandak system are Seti Gandak, Marsiandi and Trisuli. The combined Gandak has cut across the lower hills and flows through the thickly forested Chitwan valley to enter the plains.

Kosi

The Kosi river system drains eastern Nepal. It is also known as the Sapt Kosi in its upper reaches as it is formed by seven main rivers, e.g. Sun Kosi, Indrawati, Bhola Kosi, Dudh Kosi, Arun, Barun and Tamur. The Dudh Kosi drains Mt. Everest massif, the highest group of peaks in the world. The river Arun rises in the trans-Himalayan zone of Tibet. The river Barun rises from Barun glacier at the base of Makalu, the fourth highest peak in the world. The river Tamur is the eastern tributary of the river Kosi. It rises from snow on the western flank of the Kanchenjunga group of peaks known as the Kumbakaran Himal. The river Kosi then flows towards south to emerge from the hills through the Chatra gorge.

5.1.3 Brahmaputra River System

The headwaters of Brahmaputra (Yalu Zangbu or Tsang po) river system are immediately adjacent to those of the Indus river, north of the Greater Himalayas. The river rises in the great glacier of the Kailash range, just south of the Lake Konggyu Tsho in Tibet. It flows in easterly direction for about 1700 kms. Draining the southern part of Tibet (Fig 5.1.3.1). Thereafter it abruptly turns southward at the base of the Namche Barwa peaks and enters India as the Brahmaputra flowing out on the plains of Arunachal Pradesh and Assam. The river joins the Ganges in southern Bangladesh and enters the Bay of Bengal. The Brahmaputra ranks amongst the longest rivers of the world. 2880 km long river traverses its first 1625 km in Tibet, 918 km in India and the remaining 337 km in Bangladesh. The average width of Brahmaputra valley is about 86 km of which the river itself occupies 15-19 km. The river forms almost a trough receiving flows of its tributaries both from north and south.

The river Brahmaputra is joined by a large number of tributaries emanating from the eastern Himalayas and from the hills of Assam, Meghalaya, Manipur, Tripura, Mizoram and Nagaland. The main rivers draining into the Brahmputra from the eastern Himalayas are Teesta and Manas river systems.

Teesta

The Teesta river system drains the Darjeeling hills and Sikkim. It rises from Zemu glacier on the eastern slopes of the Kanchenjunga massif. Thereafter it flows along southerly course before merging with the Brahmaputra in the plains of Bengal. The Teesta falls sharply within a few kilometres of its source in the Zemu glacier dropping to an elevation of about

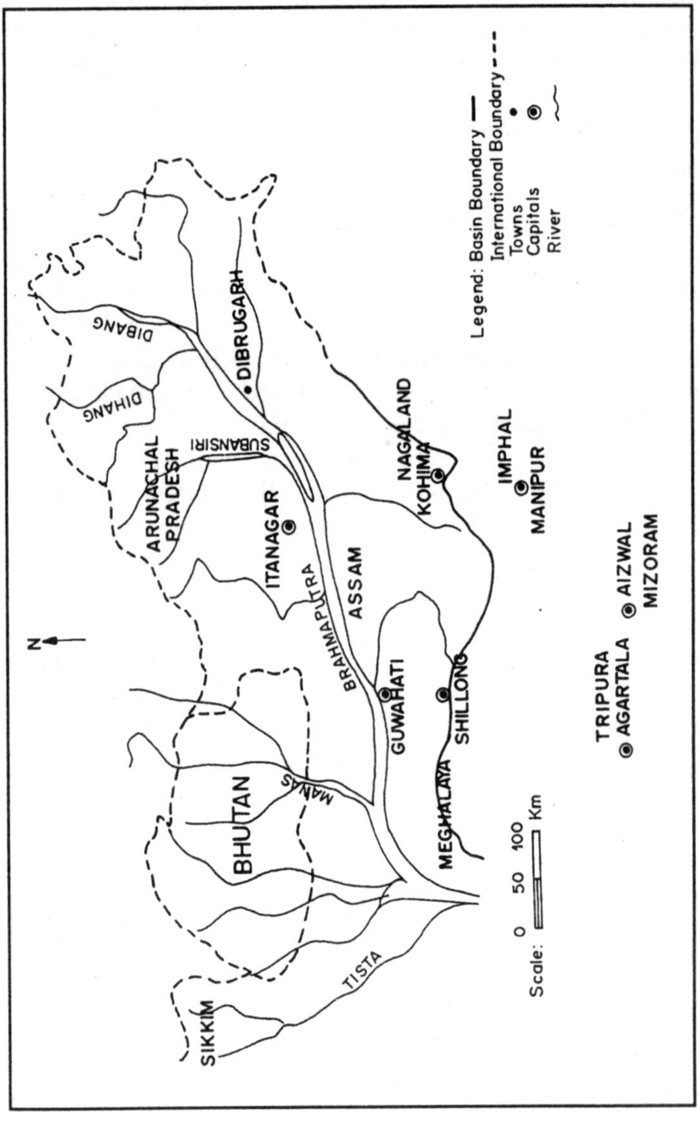

Fig. 5.1.3.1: Brahmaputra River Basin with Tributaries.

1000 m near Singhik and Mangan in Central Sikkim. The deep gorge of the river Teesta divides the hills of Darjeeling into two parts, i.e. Darjeeling Tiger hills range in the west and the Kalimpong hills to the east. Lhonak, Lachung and Rangit are the main tributaries of Teesta.

Manas

The Manas river system drains parts of central and eastern Bhutan. The main river rises in the southern slopes of the main or great Himalayan range in the NE part of Bhutan. One of its streams also rises in Tibet and cuts across the main Himalaya. It flows in a SSE direction and then makes a gradual turn towards SE near Tashi Ganga. The main tributaries of Manas are Mangde Chhu, Chamke and Kurd rivers.

5.2 Drainage Characteristics

In mountain systems, the most critical factors controlling drainage characteristics are topography and meteorology. The topographic factors of local relief, slope angle, and aspect influence the timing; volume and spatial variability of water and energy in a mountainous terrain. In mountains, from meteorological point of view, there is no regional climate as such, but rather a mosaic of local "topoclimates," determined by variations in slope angle, aspect, and relative altitude. (Alford, 1992). These interactions determine the drainage characteristics of mountain watersheds. The main drainage of Himalayan mountain system is essentially antecedent (prior to uplift) with rivers such as Sutlej, Indus and Brahmaputra commencing on the northern flanks of the mountain, flow to the south through deep gorges.

In high mountain watersheds, the runoff contributions do not depend on direct precipitation inputs but on the interaction of precipitation with thermodynamical characteristics of the environment (Bahadur, 2000). Variations in energy availability lead to fluctuations in melting of snow, ice and the production of melt-water. Because of the thermal threshold, the snow and ice masses are prevented from entering the liquid phase until the crucial melting temperature has been attained. Seasonal variation in the form of precipitation (from winter snowfall to summer rain) and of energy supply usually peaking to a summer maximum, produce strong seasonality of hydrological activity in mountains, influencing quantity, quality, and timing of drainage. Temporal variation throughout an annual cycle is accompanied by spatial variability of energy availability and precipitation quantity and form, and hence water storage (as snow and ice) within a catchment area.

Precipitation input enters temporary storage, if solid or the drainage, if liquid. When rainfall occurs on the surface of snow and ice, meltwater is released which enters the drainage system. In the first stage of the drainage process, the rainfall or meltwater passes through or across the medium forming the terrain surface. In the network of surface or subsurface flow, rain and meltwaters are routed to the catchment outlet to form runoff.

In general, hydrological system of snow and glacier covered basin is complex. Snowpack can act as temporary storage of rainfall but the snow itself undergoes continuous metamorphism to firn and subsequently glacier ice. A further complication is that the initial drainage process for leading liquid water from the site of precipitation is often related to movement of water through the mass of the stored solid water (either snow pack or glacier ice). Thus the movement of liquid water through various forms of its own solid phase, in diffuse and concentrated flow is of critical importance for the response characteristics of high mountain areas. It may be noted that a glacier is a persistent, stable and fixed location feature of a basin throughout an annual hydrological cycle. But the snow-cover which covers both ice and ice-free areas in winter months, is transitory and variable in areal content and spatial distribution. Snowpack acts as a source of meltwater production and a substrate through which meltwater and rainfall have to drain. In this role, variable spatial distribution of snowcover during the year, determines the changing proportion of the basin area under glacier ice which is available for meltwater production and which acts also as a drainage substrate. Therefore, through time, the quantity of liquid water drainage reflects not only energy supply variation but also changing source areas exposed to melting. Exposure is associated with the changing locations and proportions of catchment area contributing liquid water to runoff.

The high mountain hydrological system could have glacier-covered and ice-free zones. In some catchments of high relief, permanent ice may be absent, as every year all the accumulated snow melts during summer. In other mountain basins, some snow will persist from one year to the next, at higher elevations nourishing glacier-cover over a portion of basin area. In the remainder, almost all the ice will melt leaving a few residual snow patches which may persist through nival (ice-free) and partially glacierized basins, are that ice melt provides a source of runoff throughout summer. From the time at which snow first clears from the glacier tongue, and that the lower albedo of ice, by comparison with snow leads to higher melt rates. Fig. 5.2.1 shows various hyrological elements of a typical Himalayan valley glacier. The precipitation vary between mountain ranges within mountain ranges, seasonally and one year to another, type of precipitation also vary; the freezing level is not constant and thus the proportions of rain and snow from place to place and through time vary considerably (Young and Hewitt,1992).

Due to greater precipitation inputs and reduced evaporation in high mountainous regions, specific runoff is generally higher than in surrounding plains. With the increasing elevation of basin, specific runoff is usually enlarged. However, the distinctive characteristics of the high mountain runoff relate to the timing of flows. Because some of the precipitation accumulates in snowpack during winter, at temperatures below the melting point of 0°C, much runoff is delayed from the time of precipitation until later in the year. Not only is the solid precipitation retained, but rain falling over a snow cover is absorbed and retarded. Discharge flows of streams and rivers in mountains are therefore concentrated in spring and summer months when water is released by melting from snow and ice. By contrast, winter discharges are lower or negligible, according to the duration of temperature below freezing, controlled by the frequency of fluctuation of the freezing level. Mountain runoff largely reflects change in heat energy available for melting. Whether the melt is derived from seasonal snowcover alone or from a basin also containing

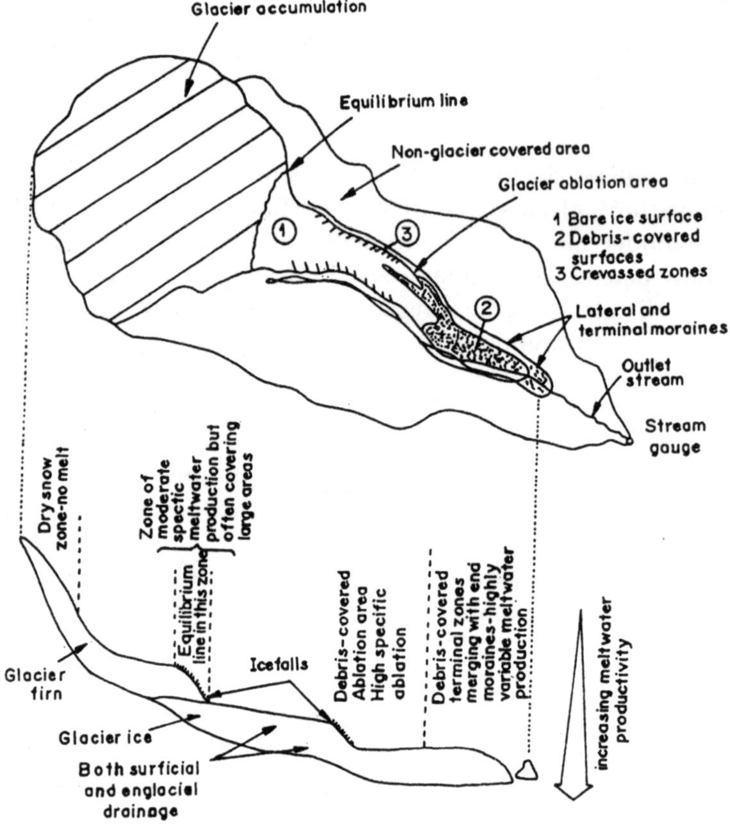

Fig. 5.2.1: Hydrological Elements of a Typical Himalayan Glacier.

perennial snow and glacier ice, runoff is a hydrological response to variation in thermal conditions.

As an illustration, the discharge hydrograph, rainfall, and air temperature record on Dokriani glacier, Garhwal Himalaya during the ablation season of 1994 is shown in Fig 5.2.2. The pattern of runoff first follow the snowmelt as the transient snowline receded from the non-glacier area where the recession process is uniform and slow. As the summer advances, the transient snowline receded well up to the equilibrium line this results in discharge peak by the end of June and early July. However, from mid-July to September thick monsoon cloud cover over the entire Garhwal Himalaya cuts off the radiation and the melting of ice substantially reduced and the discharges were mainly sustained by the runoff generated by the monsoonal rain. As a consequence of radiation interception by the cloud cover the resulting hydrograph shows a very distinct bite during monsoonal period between July and September (Collins and Hasnain, 1995). From mid-September to November, the discharge gradually reduces as the atmospheric temperature steadily falls.

The monitoring of discharge from a glacierized basin is difficult but important task for the following reasons:

- discharge is highly variable with strongly marked seasonal variations as well as pronounced diurnal changes during the melt season
- high kinetic energy of water flow may transport a rather high material load, ranging from fine-grained sediment to rocks and boulders, which by their temporary disposition often change the river bed geometry.
- sediment transport causes heavy abrasion and suffering of the measurement structure, and harsh climatic conditions are critical factors in measuring system.
- long term climatic fluctuation influence glacier behaviour, which in turn is reflected in runoff values.

In addition to storage as snow and ice, liquid water may be stored in glaciers and in marginal lakes against glaciers. An outstanding characteristics of glacial runoff is the occasional outburst flood created by sudden drainage of large quantities of water stored in englacial, subglacial and ice-marginal sites. Even small water bodies or lakes can produce very large floods because of the abrupt release of water. Some lakes may dump several times each year, others annually, and larger lakes every two or five years.

High mountains hydrology is concerned with problems rather different from those of low elevation areas. Many conventional hydrological production techniques cannot be used, usual precipitation–runoff models are inappropriate. A unique set of physical hydrological processes is associated with snow and ice in mountains. Investigations of these processes require information about the properties of snow and ice and the behaviour of glaciers in addition to an understanding of the components of the usual liquid water equation.

Average annual discharges of over 1200 km^3 the world's largest water drainage system of Indus, Ganga and Brahmaputra rivers is given in Table 5.2.1. The specific annual discharge of water (volume per unit area per year) is also given for mountain watersheds and the river basin as a whole.

From the table 5.2.1, it is seen that the volume of water draining increases from Indus to Brahmaputra. Further higher specific annual discharges for the mountain watersheds is due

Table 5.2.1: Average Drainage Characteristics of Himalayan Rivers

Sl. No	*River with measurement Station*	*Average annual discharge km³/yr*	*Specific annual discharge of water (mm)*	
			Mountainous Watershed	*River Basin as a whole*
1.	Indus near Arabian sea	207.8	460	163
2.	Ganga at Hardinge Bridge	494.3	975	473
3.	Brahmaputra at Bahadurabad	510.4	1039	922

Source: Alford, 1992.

to orographic lifting of moisture. This orographic contribution is higher being about three times mountain watershed as compared to the basin as a whole for arid Indus due to major

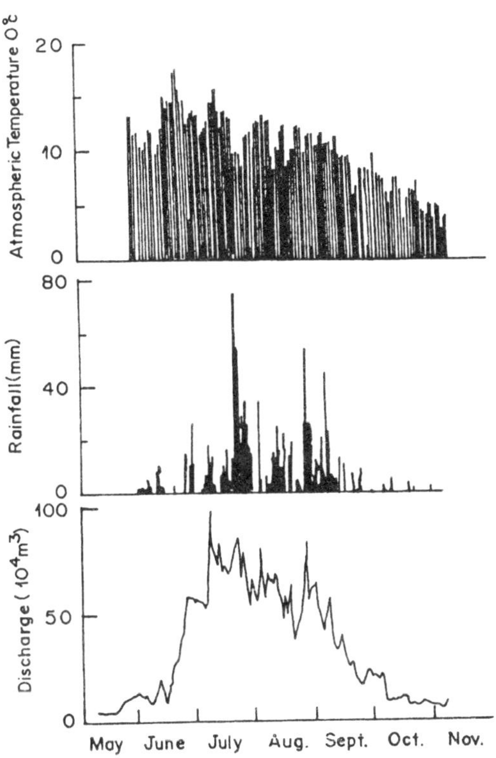

Fig. 5.2.2: Seasonal Discharge Hydrograph, Daily Mean Rainfall and Daily Mean Atmospheric Temperature of Dokriani Glacier, Ganga Headwaters, Garhwal Himalayas.

precipitation occurring in winter months due to westerly disturbances. It is least for humid and per-humid Brahmaputra being only about 10 per cent higher for mountain watershed as compared to whole river basin due to excessive and extensive summer monsoon precipitation. It is estimated that the snow and glacier melt contributions are about 50 per cent in the western Indus basin which gets reduced to about 10 per cent for eastern Brahmaputra. Another estimate shows that the specific runoff in the Himalayas is at a maximum in the altitude range 1500 to 3500 m and the volume of this water is about 515 km^3/yr from the upper mountains (Alford, 1992).

5.3 Amount of Snow and Ice Melt Contributions

Three main types of snow melting are distinguished in the Hindu-kush advective i.e. very slow melting when the sky as overcast or the weather reaction is dull; *radiation and advection*, when the most intensive melting at the rate up to 100 mm/day occurs, these being few clouds in the sky and the *radiation type* when melting is applicable on a cold sunny day (Kotlyakov and Lebedeva, 1998)

By contrast, in the high eastern Himalayas, an advection and radiation type of melting occurs during the monsoon period, when snowmelting at the role of 20 mm/d continues daily in conditions of contininous and variable cloudiness and snowfalls. When the monsoon is over, a radiation type of very snow melting is established there. The entire high Himalayan region as characterised by snow and ice penitents formed during radiation type melting under an extreme manifestation of weather pecularities: a massive influx of direct solar radiation, sub-zero air temperatures, and the low pressure of water vapour. In the course of melting, a strong effective radiation and a heavy consumption of heat for evaporation blocks solar radiation consumption for the melting of the horizontal surface. Only local micro-segments exposed normally to the rays of the mid-day sun receive some surplus radiation heat, thanks to which depressions on an even surface melt through increasingly, ultimately producing snow and ice penitents.

A maximum ablation of 400 mm/yr is noted on the NW slope of the Hindu-kush and is attributed to good humidification from Mediterranean cyclones. Ablation of glaciers along the outer rim of high Eastern Himalayas in equally strong. The maximum ablation on the windward macro-slope equals 2000 mm/yr. By and large, on the leeward macroslope of the Hindu-kush and the Himalayas, ablation amounts to ~500 mm/yr, which is four to eight times less than on the windward side.

The ratio of evaporation to glacier ice mass wasted does not exceed 5 per cent. In the interior regions, the value is lower but the melting component in ablation goes upto 20%. the only exception is in its eastern part where summer evaporation gives way to condensation due to saturating humidity of the monsoon air masses very active there. For the rest of the year, this part, too, is dommated by evaporation.

The amount of snow and ice melt contributions in Himalayas vary from year to year depending on the amount of precipitation at high altitudes and the prevalent environmental conditions during the melt season. On micro scale, there is great temporal and spatial variability depending on the location and orientation of the river basins. On the mesoscale — the scale of individual sub-basins or topographic elements within a sub-basin, such as altitudinal belts — hydrological environments range from glaciers or cold dry deserts to hot wet, low-latitude sub-tropical forests. Sometimes these contrasting environments exist within a few tens of kilometres from one another, often within a sub-basin. These contrasting environments produce a situation in which hydrological diversity of a catchment basin cannot be inferred directly from measurements at individual hydrometric stations. On the macro-scale hydrological environment of three major river basins, i.e. the Indus, the Ganga and the Brahmaputra river, provide only a limited insight into the great diversity of such environments present on the meso-scale within individual mountain tributaries to these rivers. Specific annual runoff (i.e. depth of water produced per unit area) for tributaries of upper Indus river vary from 210-910 mm; for tributaries of upper Ganga from 813-2818 mm while those of upper Brahmaputra from 119-2587 mm. These figures contain snowmelt, glacier melt and also runoff from monsoon rainfall. The rainfall contribution is minimum for Indus and maximum for Brahmaputra while the snow and glacier melt contributions are maximum (more than 50%) for arid Indus and it reduces for Ganga and is minimum (about 10%) for the annual flows for the Brahmaputra.

In absence of records of areal snow cover, its depth and water equivalent, it is difficult to assess the snowmelt runoff.

The area of maximum snowcover over Himalayan region could vary from over a 10^6 km^2 but less than 2×10^6 km^2. Assuming an average areal precipitation of 500 mm/yr over the whole region, an average snowmelt runoff could range from 500 to 1000 km^3/yr. Further the total area of glaciers in the Himalayan region is about 100,000 km^2. The annual water yield from glacier melt ranges from 3 to 6 m/yr. This gives a glacier melt contribution to an average value of 300 to 600 km^3/yr. Thus the total snowmelt and glacier melt contribution from Himalayan snow and ice work out to be (500 + 300) to (1000 + 600) km^3/yr, i.e. 800 to 1600 km^3/yr. At least 50 per cent of these meltwater contributions, i.e. 400 to 800 km^3/yr drain down the Indus-Ganga-Brahmaputra river system as against very conservative estimates of 200 km^3/yr (Upadhaya, 1995). The above estimate of 400 to 800 km^3/yr is comparable to the specific runoff of 515 km^3/yr from upper Himalayan river catchments in an altitude belt of considerable human activity, i.e. 1500 to 3500 m, as evaluated earlier (Alford, 1992).

5.4 Effect on Floods and Droughts

Surprisingly, the amount of precipitation in the form of snow has an inverse effect on the amount of runoff. Precipitation strongly affects the reflectivity; new snow is highly reflective so that it absorbs less heat and therefore melts more slowly. Old snow, firn and glacier ice have a low reflectivity. Thus, the greater the precipitation in the form of snow, the longer the glacier is covered by a highly reflective snow, and the less the runoff. A decreased amount of snowfall leads to a low-reflectivity surface of glacier ice, being exposed to solar radiation longer than usual, producing greater meltwater and increased runoff. Thus runoff from mountain glaciers, is naturally buffered or regulated in a way beneficial to mankind, producing increased runoff at times of high temperature and deficient precipitation (i.e. drought conditions) and storing water when the need for it is lower during a flood year. For a long period record for the North Cascades, Washington, USA, variability of summer runoff in different drainage basins is studied as a function of the amount of glacier cover. Variability is expressed as the co-efficient of variation, which is the variance of the total runoff for a given summer (May to September) divided by the mean summer runoff over a long period of record. It is noted that the addition of glacier cover amounting to only 10 percent of the basin area, causes the stream flow variability to drop to half the non-glaciarized value (Meier, 1982). Thus demonstrating a great regulation to flows of glacialed streams (Fig. 5.4.1).

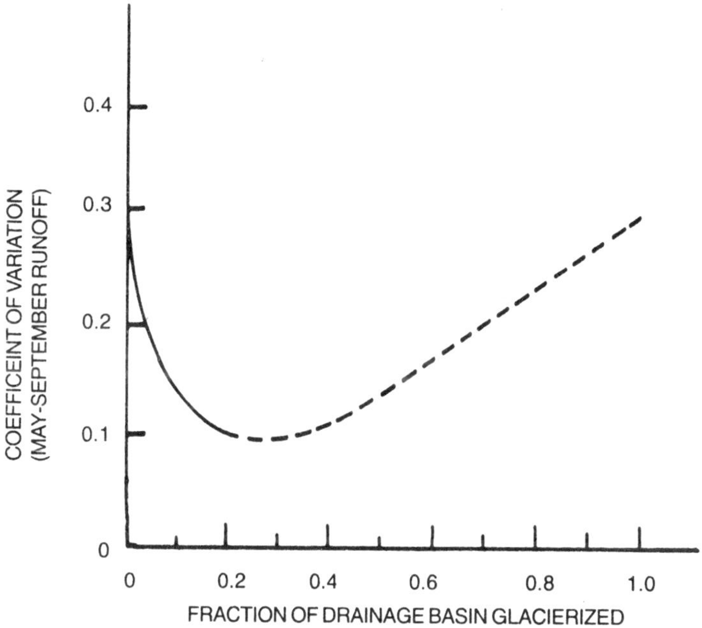

Fig. 5.4.1: Moderation of Water Discharges of Glaciated Streams.

From the above, it is clear that the hydrological behaviour of snow and glaciers has a great moderating influence on streams discharge as they release more water in an atmospheric drought year due to excessive heat from dry and hot air and less water in a meteorological flood year due to low temperature and moist conditions (more cloudy conditions).

5.5 Man-made Water Reservoirs

About 60 per cent of Indian water resources are available in the Himalayan river basins of Indus, Ganga and Brahmaputra. There is, however, a great variability in water availability in space and time throughout the mountain region. Hence, building up of storage reservoirs becomes essential for a variety of uses such as irrigation, hydropower generation, domestic and industrial requirements, flood moderation etc.

Storage capacities evaluated till 1993 is given in Table 5.5.1 showing that only 14.2 per cent of the average flow of water has so far been planned to be stored in the Himalayan region. Additional storages are needed for constantly assured water supplies to the inhabitants of the mountain region.

Table 5.5.1: Water Resource Storage and Projects in Himalayan Region

(*in km*^{3})

River System	*Average Annual Flow*	*Live Storage Capacities*				*Percentage of likely storage to average annual Flow*
		Completed projects	*Projects under construction*	*Total*	*Projects under consideration*	
Indus	3.31	13.82	2.45	16.27	0.27	23
Ganga	525.02	37.84	17.06	54.90	29.56	16
Brahmaputra and Barak	585.60	1.09	2.40	3.49	63.35	11
Total	1183.93	52.75	21.91	74.66	93.18	14.2

Source: Central Water Commission, New Delhi, 1996.

The percentage for Indus, Ganga and Brahmaputra being 23, 16 and 11 per cent respectively leaving a substantial scope for development of water resources.

There is no major water storage reservoir on Brahmaputra and Table 5.5.2 shows the major man-made storage reservoirs on various Himalayan streams.

Table 5.5.2: Dams on Himalayan Rivers (> 100 m)

Sl. No.	*Dam*	*River*	*Height*	*Reservoir Capacity 10^6 m^3*
1.	Bhakra	Sutlej	226	9,868
2.	Pondoh	Beas	116	8,141
3.	Kalagarh	Ramganga	126	2,369
4.	Thein	Ravi	147	3,300
5.	Kothar	Kosi	155	4,080
6.	Kishan	Tons	253	2,400
7.	Tarbela	Sindhu	143	1,367
8.	Mangla	Jhelum	118	7,250
9.	Tehri*	Bhagirathi	261	3,539
10.	Lakhwar	Yamuna	192	580
			Total	51,464

*under construction

From the above, it is seen that the melt-water contribution in arid Indus river system have been utilized to a large extent while the semi-arid to humid Ganga system is the next in creating reservoirs while the water resources of per-humid Brahmaputra are yet to be exploited.

6

Soil Erosion and Sediment Transport

Soil erosion includes all processes that result in the physical lowering of the surface of the earth. Surface erosion is the loss of surfacial materials, as a result of action of falling and running water or wind, while mass wasting involves the movement of large masses of fractured bedrock or other unconsolidated materials, including soil from a slope. Sediment transport is the movement of the products of erosion through a river system. It is not possible to consider water resources management in the Hindukush Himalayas without also considering the sources of the very large volumes of sediment moving annually through the river systems of the region. The limited data of sediment load measurements on Himalayan rivers (Fig. 6.1) show that the values for the region exceed the world average by almost two orders of magnitudes (Ferguson, 1984). A central concern in the development or management of the water resources of the region is the relationship among land use, on site erosion, and sediment transported through the river systems.

Erosion and sediment transport are both processes associated with the work done by water as it moves through the terrestrial portion of the hydrological cycle. This ability to perform work is defined in terms of power — the rate at which work is performed. In simple terms, the amount of work that falling or flowing water is capable is defined by the kinetic energy of water. The energy in precipitation is related to drop size and depth of precipitation during some defined time period — the precipitation intensity. The ability of that precipitation to promote erosion will depend upon a complex inter-relationship among surface properties (e.g. infiltration rate, particle cohesion, coefficient of friction, density of vegetation cover, slope angle). Generally accepted methods to quantify these inter-relationships have yet to be developed as most of the studies on erosion remain highly site-specific and empirical.

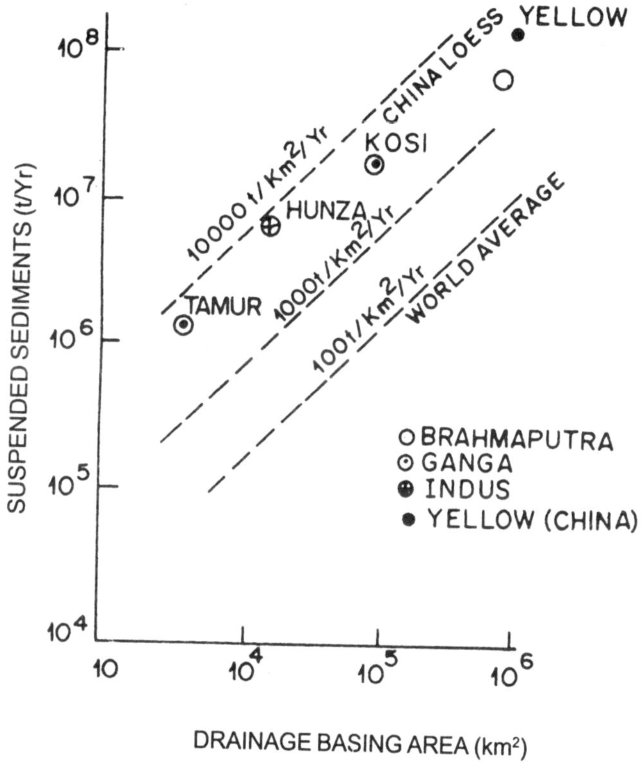

Fig. 6.1: The Sediment Load of Selected South Asian Rivers as Compared to the World's Average.

Source: Ferguson, R. (1984).

The power of flowing water is defined by the depth of flow per unit time and the angle of the slope down which the water is flowing. For water flowing through the channel, it is apparent that an increase or decrease in either stream discharge or slope angle will have the same effect on the power available to cause erosion or sediment transport. In an ideal mountain basin, these factors work in opposition. At points nearest the headwaters, slopes are steep, but stream discharge is low. With increasing distance from headwaters, slope angles decrease, but the stream discharge increases as more and more tributaries add their run off to main channel.

There are a number of natural features of the Himalayan region that result in large quantities of sediment are delivered to the rivers of the region for transport (Alford, 1992). They include :

(a) The glaciated nature of the basins,
(b) The limited natural vegetation cover,
(c) The extreme local relief,
(d) The fractured nature of the rock,
(e) The efficacy of freeze-thaw weathering cycle,
(f) The presence of easily-erodable glacier debris,
(g) The frequency and magnitude of landslides, mudflows, and avalanches that deliver sediment to the tributary channel.

The above normal characteristics of the mountain region should be seriously considered as only a few of them are amenable to modification using traditional environment management practices. One can have an idea about heavy sediment loads carried by the Himalayan river from Table 6.1.

It may be noted that the effects of erosion and sedimentation provide favourable conditions to river shifting.

Table 6.1: Sediment Yields of Selected Himalayan Rivers*

River	*Station*	*Sediment yield T/ha/year*
Indus	Kotri	5.1
Kabul	Nawshera	2.9
Chenab	Alexander Bridge	1.7
Jinsha	Batang	0.8
Jinsha	Pingshan	19.2
Lan Chang	Jivzhou	2.5
Nu Jiang	Dao Je Ba	1.6
Nian Cu	Jiangza	1.5
Lhasa	Lhasa	0.4
Yarlungtsangpo	Nuxia	1.0
Brahmaputra	In Bangladesh	14.5
Ganga	In Bangladesh	15.7
Karnali	Chisapani	67.4
Naryani	Narayanghat	56.8
Bagmati	Chobar	30.3
Kankai	Mainachuli	48.4
Kosi	Chatra	31.3
Tamur	Tribeni	82.2
Teesta	Anderson Bridge	125.1

Source: Holeman, J.N., 'Sediment Yield of Major rivers of the world', Water Resources Research Vol. No. 4, Aug. 1968; Rivers and lakes of Xizang (Tibet) Science Press, Beijing, 1984.

River Kosi has shifted by about 150 km to the west during last two centuries. The above table gives only the suspended sediments as very little reliable information is available about the quantitative aspects of bed load movement.

7

Environmental Problems of High Himalayan Region

Environmental problems of high Himalayan region are associated with its inherent fragility due to immature geology and extreme weather and climatic conditions due to intense freeze-thaw cycle on one hand and cloud-bursts on the other. We have to face the problems associated with high seismicity, climatic extremes, large snow/ice avalanches, debris flow, mud flows and debris slides, outburst from moraine dammed lakes and glacial-lake-outburst floods which occur with varying intensity and frequency throughout the mountain region, making it a high risk zone.

7.1 High Seismicity

The high seismicity (the form and status of earthquake occurrence) of the Himalayan region is rooted in its recent orogenic evolution, as a result of which Himalayas are still rising. The Himalayan province on the whole is geodynamically very active — it is prone to violent crustal movements causing earthquakes (Valdiya, 1998).

The snapping and attendant slipping of rocks on faults and thrusts have given rise to frequent earthquakes. The magnitude of seismic events has varied from moderate to high. Four major earthquakes of magnitude higher than 8 had occurred in the region. Shillong (1897) magnitude 8.7; Kangra (1905) magnitude 8.4; Bihar (1934) magnitude 8.1; and Assam

(1950) magnitude 8.7. Most of the damaging earthquakes in Himalayas have shallow focal length while others ranging from a few km to 100 km (Srivastava, 1998).

The southern most unit of the mountain system called the Shivalik sediments form low elevation hills that rise in front of the Indo-Gangetic plains. These outer Himalayas are separated from the lesser Himalayas in the north by Main Boundary Thrust (MBT). Further, north, the Main Central Thrust (MCT) separates the lesser Himalayas from the Great Himalayas. These two thrusts MBT and MCT dip steeply at 30° to 40° respectively near the earth's surface but flatten out at depth. These thrusts/faults are responsible for distribution of seismicity throughout the Himalayas with a fairly high concentration of seismicity between MBT and MCT. There are seismic gaps where major earthquake activity has not occurred during past 100 years and are considered as high potential areas for occurrence of future earthquakes.

It is feared that a large dam may induce occurrence of a large earthquake. So far nine dams have already been impounded with water with no evidence of increase of seismicity. On the other hand, a decrease in seismicity with a distance of 100 km from the site after initial filling of Tarbela reservoir in Pakistan has been observed (Jacob *et al.*, 1979).

However, the threat of great earthquakes has to be taken care by appropriate earthquake-resistant designs.

The 20 October 1991 earthquake of magnitude 6.6 on the Richter scale in the Uttarkashi-Tehri region of Garhwal (Central Himalaya) has again brought havocs and perils in areas of active faults. The earthquake which violently shook a large part of north, central India, originated as a result of slipping on segments of faults in the active zone of the Main Central Thrust (MCT). The Garhwal region has witnessed several earthquakes as listed below (Table 7.1.1) by the School of Research and Training in Earthquake Engineering, University of Roorkee, U.P., giving us an idea of seismicity of this central Himalayan region.

Table 7.1.1: Earthquakes in Garhwal Region*

Date	*Epicentre*		*Magnitude on Richter Scale*	*Distance from Badrinath (km)*
	Lat°N	*Long°E*		
1816 May 26	30	80	6.5	94
1843 April 11	30	80	50	94
1902 June 16	31	79	6.0	52
1906 June 13	31	79	5.56	52
1908 Dec 11	31	79	5.0	52
1945 June 4	30.3	80.0	6.0	70
1947 June 19	31.2	79.9	5.6	75
1948 May 5	30.5	78.5	4.0	90
1949 Feb 5	31.2	79.9	5.6	75
1958 Dec 28	30.01	79.94	6.25	90
1958 Dec 31	30.09	79.86	6	75
1962 July 13	30.5	79.6	5.6	30
1963 Nov 27	30.8	79.01	5.1	30
1968 May 31	29.9	80.0	5.1	100
1969 June 22	30.8	79.3	5.8	18
1970 May 13	30.6	78.7	4.5	70
1991 Oct 20	30.51	78.29	6.6	70

**Source* : Bhatt, 1992.

The highest earthquake in the area was of the magnitude of 6.6 on the Richter scale at a distance of 35-40 km from 260.5m high Tehri dam, which has been a controversial project for over last three decades. The dam is located 4 km south of active Shrinagar Thrust, cutting through a region which is likely to be rocked by a future earthquake of magnitude of 8 or higher. However, no damage to the dam foundation was noticed after 1991 Uttarkashi earthquake. Surfacial cracks to the structure were not observed anywhere in the dam area and the project authorities maintained that the design features ensure the safety and the integrity of the reservoir.

7.2 Climatic Extremes

The Himalayan region is characterized by extreme cold and arid winter conditions in the West to warm and humid summer conditions in the east with alterations of dry and moist conditions in wide range of altitudes having different micro-climates. However, the understanding of climatic extremes such as cloud bursts, hail storms, droughts etc., which play a crucial role as an environmental constraint at micro levels, is not well understood. What is worse, due to inaccessibility, proper recordings and quantification of these events, it is impossible to predict these events with any confidence. Cloud bursts cause heavy damage and slope failures while hailstorms or droughts inflict great losses to agriculture.

Cloudbursts occur due to heavy rain-shower due to intense convective activity and occurrence of thunderstorms. These storms are only 15 to 30 km across and generally escape observation from widely established meteorological network. We need closer observation network for better definition of storms for predictive purpose. Table 7.2.1 shows the most significant climatic extremes of heavy rainfall, which are the root cause of devastating floods. Maximum observed 24 hour rainfall totals are given which follow the same pattern in terms of spatial variation of intensity.

Table 7.2.1: Maximum observed 24-hour Rainfall at Selected Mountain Stations*

Sl. No.	*Station*	*24 hour rainfall (mm)*
1.	Cherrapunji	1036
2.	Dibrugarh	745
3.	Darjeeling	640
4.	Almora	221
5.	Dehradun	152

**Source*: Bruiznzeel, L.A., Bremmer, C.N., ICIMOD Occasional paper No. 11, Kathmandu, 1989.

The cloudbursts is one of the major natural hazards in Himalayas, occurring amid extreme hydrometeorological conditions. It strikes at random and at hightening speed, generally lasts for a limited time, and leaves behind a trail of devastation.

The debris flow of colluvium or forecut deposit along gullies take place at a tremondous velocity, ranging between 1 to 10. Moreover, the viscosity of the water increases rapidly with the increase in sediment concentration. The fine particles of mud become suspended in water because of the eddy currents of rapid debris flow. Consequently, the unit weight of the pore fluid from 1 to 1.6 g/cc depending upon the flow velocity and clay content. Because of the increased velocity and viscosity of the water the capacity so carry boulders increases by the abnormal drag, dragging heavy boulders along the gullies.

7.3 Snow and Ice Avalanches

Snowfall in great Himalayan range varies from 12 to 18 m in winter months (November to April) in the western region making the region more vulnerable to snow avalanches (Upadhaya, 1995). Winter snow is smaller in eastern region with a high treeline, which restricts the avalanche activity and thus the environmental hazards.

The great Himalayan range experiences maximum number of avalanches of all types ranging from loose snow avalanche during early winter to wet slab avalanche during March-April.

The avalanches occur whenever deep snow accumulates on slopes exceeding a crucial inclination. Avalanche activity is a function of structural weakness of the snow cover over the ground and the prevailing meteorological conditions. Due to excessive snowfall, steep slopes of the mountains, strong winds and tectonic activities in Himalayas, snow avalanches occur in varying intensity and frequency. Most of Himalayan glaciers are fed by snow avalanches from high slopy terrain in the glacial region. Sometimes contributions from these snow avalanches compare the direct snow precipitation inputs to the glacier system. Avalanched snow converts to glacier ice more quickly than the snow precipitated in the higher elevation, thus enhancing the glacier ice activity of the Himalayan region.

Study of geomorphological characteristics such as slope, aspect, terrain configuration along with snow density, settlement, glide etc. are of vital significance for forecasting avalanches. Potential sites are normally studied in details integrating the available data for hazard evolution and forecast.

Snow Avalanche Study Establishment (SASE), Manali is working on avalanche control structures, e.g. design of formation zone control structures like snow nets, snow rakes and snow bridges has been executed for a site in Jammu and Kashmir. Similarly design of snow gallery, diversionary dams and moulds have been made along Manali-Leh road. We may recognize the role of the natural forests as a supporting structure over the costly engineering measures and artificial structures.

Ice avalanche is a sudden, rapid, downslope movement of ice following its detachment from the terminus of a glacier. Conditions favourable for ice avalanching are created when the terminus of a glacier retreats up a steep slope. Glacier avalanches are common in mountain areas and are hazardous. When the avalanching ice plunges into a moraine — dammed lake, it can generate waves that overtop the dam, which may trigger a catastrophic outburst (Evans and Clague, 1992). Glacier avalanches usually occur during the summer months and result from a destruction of tensile strength in the ice mass through progressive fragmentation associated with crevasse development, the melting of parts of the glacier may be frozen to the substrate, and the reduction of frictional resistance at the ice/rock interface due to increased water pressure. A few years back, an ice avalanche in Beas destroyed the equipments building and other infrastructure developed at the campus of SASE, Manali, which had to be shiffed near Chandigarh.

7.4 Debris Flows, Mudflows and Debris Slides

Debris flows, mudflows and debris slides are all mass movement processes and could be included in one word, i.e. mass wasting (Gerrard, 1990). Mass wasting is a widespread constraint faced in all parts of the Himalayas. It describes a variety of processes through which large masses of earth material move downhill under gravity, both in slow creeping mode or as rapid landslides.

Mass wasting is the most frequent and widely dispersed manifestation of the fragility of the Himalayan terrain and is basically a natural process. Human activities can substantially aggravate its impact and scale. Most important natural factors are:

(i) Steep slope with high relative relief,
(ii) Seismicity,
(iii) Groundwater flows accentuating landslips,
(iv) Cloudbursts and intense rainfall events,
(v) Nature of the rocks (soft sedimentary, foliated metamorphic or fractured igneous),
(vi) Toe undercutting by torrents and floods.

The above factors are aggravated by anthropogenic activity such as:

(i) Loss of forest cover,
(ii) Extension of agriculture into steep slopes,
(iii) Open-cast mining without environmental control,
(iv) Road built without regard for geological factors.

The occurrence of landslides in mountains cannot be prevented in all cases but the damage can be reduced by taking effective control and preparedness measures for disaster management. Land hazard zoning characteristics, e.g. proximity to fault, slope angle, dip slope retention, relative relief, height, nearness to ridge tops, lithology and land-cover is helpful in making decisions for development projects. Large catastrophic landslides also occur specially on the slopes of the main glacially over-steepened valleys and the major river gorges. These hazards are relatively rare but when they occur, they present a great risk to the population.

7.5 Outbursts from Moraine-Dammed Lakes

Moraine-dammed lakes are found in high mountains close to existing glaciers. They are formed when glaciers retreated from moraines built during the Little Ice Age and where the moraines dammed glacial streams (Evans and Clague, 1992). Moraine dams are susceptible to failure because they are steep-sided, have relatively low width-height ratios, and consist of poorly sorted loose sediments. In addition, these dams and the lakes behind them commonly occur immediately downslope from steep slopes that are prone to great avalanches and rock falls. Moraine dams generally fail by overtopping and incision. The triggering event is most frequently a glacier avalanche from the toe of the retreating glacier which generates waves that overtop the dam. Melting of ice cores and piping are other reported failure mechanisms.

In Himalayan region, of some 3000 to 5000 m altitude over a distance of 10 to 30 km, from the main river valleys, are reflected in the steep fall of glaciers that cause these problems. On the other hand, the trunk streams of Himalayan rivers are deeply incised and fall relatively gently. Steep glacierized valley walls and tributary valleys may thus cut off extensive drainage areas. High relief, steep average fall and large climatic gradient between upper and lower reaches promote vigorous activity of ice masses in the region. This is reflected in those cases where the glacier movement is of the order of 150 to 250 m/yr. The sudden drainage of dammed lakes represents a great hydrological hazard.

7.6 Glacier Lake Outburst Floods (GLOFs)

Glacier-dammed lakes are found mainly at the margins of valley glaciers, although some occur within or beneath cirque and valley glaciers and mountain ice caps. Some of the largest lakes are situated in main valleys dammed by tributary glaciers and at the mouths of tributary valleys blocked by trunk glaciers. These lakes may drain suddenly and rapidly by the formation and enlargement of subglacial and englacial tunnels, and occasionally by overtopping: the resulting flood is termed a jokulhlaup (Evans and Clague, 1992) or glacier lake outburst flood (GLOF).

Thus GLOFs are due to sudden outburst of water from glaciers or glacier-dammed lakes. In glaciated areas, these outbursts occur frequently, at periodical intervals and generally have no direct relationship with the meteorological conditions. These are primarily controlled by hydraulic considerations. A GLOF could be large enough to destroy roads, bridges or other structures and may be truly catastrophic in timing and release of amount of water and sediment quantity for the destruction of the downstream environment. GLOF may involve several types of impounded water bodies. These may be large visible lakes in the lateral streams and valleys or lakes in the main valley dam by ice from a tributary valley or the water may be entirely below the surface of ice and not visible. These floods are normally periodical but the periodicity may vary from 1 to 10 years or even more depending on the location. The resulting hydrograph from GLOF resemble normal storm unit hydrographs except that the time scale is reversed. The flow starts at a low rate and increases exponentially until a peak is reached then the discharge drops off abruptly. GLOFs are poorly documented but observed to create floods raising water levels up to 100 m (Bahadur, 1981).

Some of the world's largest documented historical GLOFs occurred in Karakoram Himalayas.

The damming of upper Shyok river by Chongkumdan glacier formed a lake (Volume - 1.4×10^9 m^3). A sudden outburst from this lake occurred in 1929 and the flood wave travelled down the Shyok river into Indus creating a rise of 8m, 740 km. downstream from the ice dam.

In Nepal, thirteen GLOFs are observed since 1960 for 30 years duration giving a very high frequency of natural hazard i.e. more than once in every three years (Yamada, 1992).

In another investigation, the maximum GLOF discharge at the upper Arun damsite is about three times the spillway design flood (1000 year flood) selected for the concrete structure of the run-of-river dam. Results of flood flow analyses (Meon and Schrooz, 1992) for upper Arun dam site are given below (Table 7.5.1).

Table 7.5.1: Flood Flow Analyses for Upper Arun Dam Site*

Type of Flood	*Maximum Discharge (m^3/s)*
GLOF	6300
PMF	4400
1000 year flood	2100
100 year flood	1800

**Source*: Meon, G and Schwarz, W., 1993.

It is considered necessary to have GLOF analysis for design flood of projects in glacier dominated watersheds. Unfortunately, little is known of the exact mechanics of outbreak of Jokulhlaups.

8

Monitoring for Protection and Conservation

The Himalayas is indeed a precious gift from nature which has moulded life, culture and history of our subcontinent since ages. Some of the advertent and inadvertent anthropogenic operations, viz. felling of trees, road blasting, construction of dams, reservoirs and hydroelectric projects, Jhum cultivation etc. have injured the pristine environment increasing soil erosion, frequent landslides, sinking of mountains, disruption of natural water channels, flooding of rivers, streams, irrigation canals, dams and reservoirs in plains, suggest the need for monitoring the huge ecosystem, for protection and conservation of its natural resources. Increasing tourism and mountaineering activities are further polluting the environment affecting the natural ecological balance and the purity of the mountain system.

In ancient India, emphasis was laid to learn the integrated and holistic approach towards maintaining the order of nature and preserving the environment (Misra, 1992). Our ancient culture was based on *Dharma* (duty to others), *Artha* (production), *Kama* (conception) and *Moksha* (salvation) for human existence. This is no more valid for large section of the modern civilization who are in great hurry and believe in quick solutions and we have to adopt data based approach for modelling, forecasting and management of vast natural resources of the region to the dynamic needs of our society.

Using conventional data and satellite imageries, we have to integrate the natural resources with evolving socio-economic data bases at different spatial units (village, block, district) and at different temporal domains.

8.1 Increased Pollution

The chemical composition of precipitation is strongly influenced by fossil fuel consumption as well as agricultural and cultural practices. Recent studies on snow precipitation in high mountains show high inorganic ion-burdens are associated with dusty layers in the snowpack indicating that dust from vast arid regions of Central Asia is the dominant source for major ions (Wake *et al.*, 1992). Significant increase in sulphate suggest anthropogenic emissions. Of course, these contents have a variable character depending on the local meteorological conditions.

Besides inorganic contents, some glaciers have large microbial production (algae and bacteria) on their surface which reduces albedo and accelerates glacier melting. Most of the human activities are confined to the lower 3000 m altitude in Himalayan region except a few places of pilgrimage where there are large number of visitors during summer months.

As the flood of climbers, trekkers and tourists rises, the Himalaya are getting increasingly polluted. The regulatory authorities have to ensure that the imbalances (created by visitors) in food, fuel and the ecosystem are kept within its safe carrying capacity of the natural system. Probably community based mountain tourism (equitable community benefits) implementation will restore the mountains by careful conservation efforts and enterprise development can be compatible and even mutually rewarding.

8.2 Effect of Deforestation

According to the Forest Conservation Act of 1988, 60 per cent of the land area in hilly region should be under forest cover for ensuring ecosystem and environmental stability. But dense forest cover (>40% canopy density) is presently 11.8 per cent which is not likely to increase significantly in the coming years. In other words, threat to plant species and all other living forms shall increase as their habitats shrink all over Himalayas, affecting biodiversity (Samra and Raizada, 2001) and the quality of life by disturbing the hydrological cycle.

Disturbed ecosystems provide a harsh condition for plant and microbial growth because of low organic matter, nutrient supply or compact soil structure. Deforestation increases loss of nutrients like carbon, nitrogen, phosphorous and sulphur by leakage from the ecosystem leading to reduction in soil fertility, carbon fixation and organic matter accumulation. Under these conditions, only a few non-demanding but aggressive species invade open areas and establish themselves.

A large number of domesticated animals, unlimited fodder and fuelwood collection, extraction of non-timber forest produce (NTFP) and conversion of forest land into crop production and plantation crops has led to a dramatic decline in forest cover in the Himalayan region. Since local communities exploit forests for their basic needs, their frequent intrusions into forest areas often runs into conflict with government agencies who control access over forest areas that still hold a small reservoir of species.

Himalayan biota is very much disturbed particularly in the western and central region, the eastern region being comparatively less disturbed due to the tropical humid climate and less biotic pressure. Over the past decade there has been a significant decline in the forest cover in several Himalayan states except Arunachal Pradesh, Sikkim and Uttranchal as given in Table 8.2.1.

8.3 Global Warming

A rising trend in global temperatures started from 1890 onwards. The increase in mean air temperature at earth's surface during past 100 years has been estimated to be between 0.3 and 0.6°C which resulted into a rise of sea level between 10 and 20 cm. With continued emission of greenhouse gases (CO_2,

CH_4, CFC, N_2O) from human activities, it is projected that the surface temperature may rise at a rate of 0.3 °C per decade and sea level by 2 to 4 cm per decade due to thermal expansion. Global warming affects nature and society, social impacts are felt for energy consumption, food production, health aspects and settlements.

Table 8.2.1: Comparative Changes in Dense Forest Cover in the Indian Himalayas*

States	*Forest Cover (km²)*		*Annual Growth Rate (%)*
	1987	*1997*	
North Eastern Region			
Arunachal Pradesh	51,096	54,155	0.60
Assam	18,415	15,548	–1.56
Manipur	4,670	4,937	0.57
Meghalaya	5,749	4,044	–2.97
Mizoram	2,938	4,348	4.80
Nagaland	637	3,487	–4.50
Sikkim	1,867	2,423	3.00
Tripura	340	1,819	43.50
Western Region			
Himachal Pradesh	9,908	9,560	–0.35
Jammu and Kashmir	12,970	11,020	–1.51
Uttranchal	10,506	16,622	5.82
Total	1,24,846	1,26,163	0.11

**Source*: Forest Survey of India (FSI) – 1987, 1997.

Based on the general pattern of climatic gradient along the altitude in Himalayas, it is predicted that the direct and indirect impacts of global warming will become apparent at a much faster rate in Himalayas and the effect will increase with altitude. The climatic warming may reduce the altitudinal contrasts in species diversity and population status leading to an upward shift in species colonization. Global warming will result in increased weathering of rocks and mass movements. The permanent snowline may move upward and glaciers will recede affecting the hydrological cycle. A comparative study of CO_2 and temperature effects on low and high altitude Himalayan species would enhance our ability to predict changes in ecosystem dynamics due to global warming.

With poor database on physical and biological aspects of productivity aspects, it is difficult to make precise evaluation. However, the general prediction (Purohit, 1999) can be made as follows:

(i) The growing season will shift from autumn → winter → spring→ summer.
(ii) The growing season will be extended, particularly at higher altitude.
(iii) Probably, there will be overall increase in plant productivity.
(iv) Species currently restricted to valleys will colonise higher reaches.
(v) Species with specific niche on south-facing slopes will be threatened and the south-facing slopes will become more xeric.
(vi) Broad leaf evergreen species will be particularly vulnerable and either the deciduous or the evergreen conifers, specially pines will cover larger area even on north-facing slopes.

In a study on climatic scenario for India sub-continent (Kelly and Hulme, 1993), utilizing 7 recent climatic models up to 2100 AD the results show that the annual average temperature depict, warming up over the central part-close to that affecting global mean temperature. In the North, warming will be greater compared to South where it will be not as great (likely projection of about 1°C). For Himachal Pradesh in western Himalayas, the projected temperature and precipitation changes are annual temperature rise of about 3°C; spring temperature rise about 3°C, spring temperature rise of about 7°C; summer temperature rise about 4°C, with a wide range of uncertainty. The annual precipitation change of +20 per cent; winter change + 30 per cent to –15 per cent; spring change + 30 per cent to –20 per cent and autumn change

+ 15 per cent to –15 per cent. Russian scientists estimates that the precipitation in the Hindu-kush will increase by 15 per cent in the northern part and by 10 per cent in the southern parts, while in the Himalayas, its amount will even decline (Kotlyakov & Lebedeva, 1998).

8.4 Artificial Regulation of Snow and Ice-melt

Several experiments have been performed to increase snow and ice melt runoff by artificial methods, such as by spreading coal dust, chemicals (rock salt, sodium chloride, calcium chloride with or without sand or gravel), heating using reflectors to concentrate solar energy or by using combustible substances. None of the chemicals are found suitable for routine use on watersheds for increasing melt-runoff due to other environmental problems like pollution, sedimentation of downstream systems.

Icing problems are experienced in various parts of the Himalayas above 3000 m due to defective drainage on roads, formation of wind crusts in the regions of high wind, compaction of falling snow by road traffic and refreezing of slush left over after mechanical clearance of road, are responsible for formation of ice layers on roads (Upadhaya, 1995). At many places, freezing of groundwater seepage also forms icing, a few cm thick. When the ice layers are treated with chemicals for de-icing the roads, the main heat exchange processes are:

(i) Eco- or endo-thermic heat from chemicals used.
(ii) Release of heat by lowering ice temperature due to addition of impurities to snow/ice.
(iii) Heat exchange at ice-air and ice ground interfaces.
(iv) Solar radiation inputs.

There are two other possibilities which need further investigation for their practical application to augment snow and ice melt (DST Report of the Study Group, 1976).

(i) It is observed that air temperatures 5 to 10 m above the snow surface range from 10 to 12°C. By artificial turbulent mixing, this warm air may be continuously brought into contact with the snow surface and thereby the ice melted.

(ii) There are geothermal areas in high Himalayas and this underground heat could be used to augment the snow and ice melt.

9

Utilisation of Modern Technologies

Under the subhead, we shall discuss aspects connected with the modelling of snow and glacier fields, utilization of remote sensing technology GIS-GPS technology, application of isotope techniques and the water resources management technology in the Himalayan region.

9.1 Modelling of Snow and Glacier Fields

Snow crystals after falling on the ground are continuously changing and getting more dense. The gradual transformation of snow to glacier ice results from changes in the constituent crystals by mutual displacement of crystals, changes in size and shape and internal deformation. The relative importance of these mechanisms changes as the density of the material increases. Ice molecule can be pictured as a regular tetrahedron with positive charges in two corners and negative charges in the other two. Each negative attracts a positive corner in a neighbouring molecule joining molecules together by a hydrogen bond. Thus, each molecule is surrounded by four other molecules in a regular tetrahedron arrangement. Ice is usually described as a visco-elastic solid — the initial part of the strain-stress graph is linear (i.e. elastic region where Hookes' law holds). Beyond the elastic limit, the strain increases more rapidly than the stress (i.e. plastic region) and the material does not revert to its original size and shape when stress is removed. The viscosity of ice ranges from .002 to 2000 (in units of 10^{12}

poises) to conclude that the ice is a non-Newtonian material. Further, the more the impurities in ice, the more readily it is recrystallized as opposed to the behaviour of metals (Pounder, 1965). With these background complexities about snow and ice, we shall discuss modelling aspects.

Simulating the dynamics of glacier requires mathematical modelling of the geophysical processes associated with a glacier. The main geophysical processes that take place on and within a glacier are densification, energy exchange, deformation and hydraulics. Diversification and energy exchange processes together transform the snow crystal to ice structure. Energy exchange, geomorphology of glacier, subglacial topography and rheological properties of ice determines the glacier dynamics which explains the mechanism of glacier flow, crevasse formation etc. Energy exchange in combination with glacier hydraulics determines the mechanism of glacier sliding, surging and deformation of sub-glacial sediments.

Modelling of flow in glaciers thus involves a number of factors which vary from glacier to glacier. A realistic approach to modelling flow in mountain glacier would be to use field data which allow a realistic formulation of the problem by taking into account the important parameters that need to be addressed. This approach will allow a better estimate of model parameters provided field measurements of these parameters are available. Some of the major problems for modelling Himalayan glaciers are:

(i) Lack of consistent and continuous database on mass balance of glaciers.
(ii) non-availability of meteorological data of the region and poor understanding of mountain meteorology.
(iii) absence of winter balance/accumulation information on these glaciers.

Keeping the primary aim of modelling, i.e. experimentation, explanation and prediction for understanding the behaviour of glaciers and their behaviour to climatic change, simple analytical models have been used elsewhere to estimate some of the important parameters related to glacial flow. However, in the real world with more realistic governing differential equations and boundary conditions, numerical solutions are the norms. Most of the Himalayan glaciers are polythermal (having cold and temperate zones) and thus both the mechanical and thermodynamic statements need to be made.

9.2 Utilisation of Remote Sensing Technology

Remote sensing technology has a special place for monitoring and evaluation of Himalayan snow and ice conditions for the inaccessible large terrain. Remote Sensing technology is defined as the technology which enables measurement of properties of an object or feature on earth's surface by an instrument that is not in direct physical contact with the object or the feature. The term was coined in early sixties to mainly name the process of acquiring terrestrial data and information with the aid of satellites put in the orbit around the earth. Later, it covered the data and photographs got with the help of aeroplanes, balloons etc. Basically, remote sensing is a sort of non-contact data acquisition system.

For snow cover and glacier studies, surface, near surface and deep subsurface features are analysed using sensors operating at different portions of electromagnetic spectrum, viz. visible, near infra-red, thermal infra-red, microwave and other wavelengths. Spectral reflectivity of snow is dependent on snow parameters, e.g. grain size and shape, impurity content, near surface water content depth and surface roughness as well as solar elevation. Freshly fallen snow has a very high reflectance in the visible wavelengths. As snow ages, the reflectivity of snow decreases in the visible and especially in the longer (near infra-red) wavelengths. This greater decrease in the near infra-red wavelength is largely due to melting and refreezing within the surface layers and to the natural addition of impurities.

Reflectivity of ice in the visible and near infra-red

wavelength varies considerably, depending on the overlying material, impurities within ice and the presence of surfacial melt-water. In general, reflectivity of glacier ice is quite low but it is higher when it is covered by snow and firn in the accumulation zone.

Remote sensed data of snow and glacier covered terrain could be used to estimate and predict melt water runoff (Dhanju, 1981). It is particularly useful in determination of the height of the transient snowline. Various experiments on different glaciers in various countries indicate that there is a linear relationship between the height of the transient snowline at the end of the summer and the glacier mass balance.

In the country, false colour composite, prepared by using bands 2, 3 and 4 of IRS–1A, LISS-II and Lansat TM sensors have been used to delineate various glacial features like glacier boundary, ice divide, accumulation area, ablation area, equilibrium line, glacial lakes and periglacial features. It has been extremely useful for the preparation of inventory of Himalayan glaciers. It is also very important tool to monitor the retreat or advance of glaciers of otherwise difficult and inaccessible terrain and help modelling efforts.

9.3 GIS–GPS Technology

Geographical Information Systems (GISs) are usually defined as computer systems for the inputs, storage, analysis and display of spatial data. GISs are tools that provide functions to digitise maps, or capture image data, to design and build data bases, to apply spatial analysis, and to display or output results as maps, reports, images and other digital products. These tools are very helpful in solving complex problems connected with the planning and management of natural resources.

GIS technology is used to manipulate objects in geographic space, and to acquire knowledge from spatial facts. Spatial theory provides a basis for GIS by bringing together fields that deal with spatial reasoning, the representation of space, and human understanding of space.

Spatial reasoning is oriented towards the deduction of spatial information from spatial facts. It deals with the framework and models for space and time, and the relationship that can be identified between objects in a spatio-temporal model of the real world.

Scientific methods for the representation of space are important for the development of data models and data structures to represent objects in spatial databases. These databases are distinguished from conventional databases by their capability to store and manage data with an extent in space and time. The human understanding of space, influenced by language and cultural background, plays an important role in the way the people design and use GIS as a tool.

Global Positioning Systems (GPSs) are a method of automatic spatial data logging originally pioneered for navigation and military use. A constellation of NAVSTAR satellites permits the use of a GPS receiver to calculate his/her position on the earth's surface to an accuracy of within one metre. Altitudes can also be recorded. A GPS logger can be connected to a PC for direct data input to a GIS. GPS technology is not suitable for data capture in large study areas. One essential prerequisite for the use of GPS-system is a clear field of view since it is necessary for the GPS receiver to fix its position from four of the NAVSTAR satellites. This situation ensures that both geographic location and height are fixed to the best accuracy. However, locking on to four satellites in a mountain terrain is often problematic when working in confined areas, e.g. deep valleys or forested slopes.

GPS have already been used with GIS to detect sites in mountain areas which are used for image classification.

The combined use of GIS, digital image processing and simulation models provide new possibilities for better environmental monitoring, forecasting and management for wider areas within a limited time. Using GIS in temporal perspective, it would be easy to develop a forecasting model and an optimisation model to achieve effective public policy.

9.4 Application of Isotope Techniques

Isotope Techniques as applied in snow and glacier hydrology have been reviewed since 1950 (Bahadur, 1983). Topics like determination of water — equivalent of snowpacks, location of ice lenses in snowpacks, regional surveys for assessment of water supply from snowmelt, determination of rates of snow accumulation (using stable isotopes, natural radioactive isotopes and fallout from nuclear detonations), differentiation of snow and glacier melt contributions to streamflows, dating of ice, delineation of flow patterns of glacier ice and resurrection of paleo-environments have been included.

Since 1977, the Glaciology Group of Physical Research Laboratory have been engaged in the applications of isotope techniques to study diverse glaciological problems on selected glaciers from Kashmir to Sikkim Himalayas (Nigampurkar, 1993). Highlights of these studies are as follows :

- Nuclear debris containing ^{95}Zr, ^{137}Cs, ^{144}Cs etc. was observed in freshly fallen snow in 1981 from the accumulation zone of the Changme-Khangpu glacier. This confirms the presence of nuclear debris produced during the Chinese atmospheric nuclear debris.
- Concentration of ^{137}Cs in snow samples collected on Chhota Shigri glacier in August 1987 confirms the evidence of Chernobyl fallout.
- Net accumulation rates of 0.7 m/yr on Changme-Khangpu-glacier and 0.55 m/yr for Chhota Shingri glacier using ^{137}Cs and ^{18}O respectively.
- $\delta^{18}O$–value in snout ice of Chhota Shigri glacier is depleted by 4‰ as compared to fresh snout values indicating that the temperatures were lower a few centuries ago.
- The observed major ions (N_a^+ K^+, Ca^{2+}, Mg^{2+}, Cl^-, SO_4^{2-} and NO_3^- in snow and surface ice collected at different altitudes 4050 to 4750 m from Chhota Shigri glacier are generally less than 1ppm.
- Flow rates of various glaciers are tabulated with other physical characteristics of investigated glaciers (Table 9.4.1):

Table 9.4.1: Physical Characteristics of investigated Himalayan Glaciers with Snout Ages and Average Flow Rates*

Sl. No	Glacier	Location	Altitude	Glacier length (km)	^{32}Si model age (a)	Past flow rate ma^{-1}	Observed flow rate ma^{-1}
1	Nehnar (Kashmir)	34°09′ N 75°31′ E	3920-4925	3.4	500	6	12
2	Chhota Shigri (HP)	32°15′N 77°31′E	4050-5000	9.0	250	28	23
3	Gorgarang (HP)	31°26′N 78°24′E	4765-5360	3.5	160	18	NA
4	Zemu (Sikkim)	27°43′N 88°17′E	4260-6000	26.0	120	200	NA
5	Changme Khangpu (Sikkim)	27°58′N 88°42′E	4850-5800	5.8	100	40	13

**Source*: Physical Research Laboratory, Ahmedabad.

Deep ice-core isotopic and chemical content from Himalayan ice will provide climatic variations for 1000 years or more as has been established (Lin *et al.*, 1995).

9.5 Water Resource Management Technology

Harnessing the 'Elixir of Life: Water' for overall development becomes a necessity particularly where there is a great variability of rainfall distribution both in space and time.

There has been a long controversy about big or small reservoirs in the country. Both are needed to meet the requirements of various sectors of human activity be it industry or agriculture or any other use for a large population. Defenders of large dam projects point out their necessity for long-term economic development especially in semi-arid areas with large and seasonal and interannual fluctuations in rainfall. In such regions, a dam mainly serves as a water reservoir but often also generates electricity. Critical views focus on the often profound social and environmental consequences accompanied by the inundation of large areas by large dams. Whole villages need to be resettled, cultural heritage and religious sites are flooded and the ecological consequences range from hindering the fish migration to threatening the flood plains downstream which often serve as food resource for people and wildlife.

Bridge building requires willingness by stakeholders to compromise and integrate new arguments into their own concept. When dam-defenders flag for the benefit of creating energy without reducing the fossil fuels and contributing to the greenhouse effect, dam-skeptics argue that the diminished runoff alters local climate. Only open discussions with honest interest in an integrative, creative solution can lead to a win-win option for all parties.

The World Commission on Dams (WCD) in its report released in November, 2000, deals with far-reaching consequences on the future of existing and proposed large dams. A dam with a height of 15 m or more from the foundation is classified as a large dam; dams between 5 and 15 m but having reservoir volume of 3 million cubic metres or more, are also large dams. There are about 45,000 such large dams in the world, with majority of them (67%) located in developing countries, including China (45%) and India (9%). WCD's major finding is that technical socio-cultural and environmental performances of most of the large dams have considerable scope for improvement. In some cases, the dams have caused more harm than good (Limaye, 2001).

In principle, the planning and construction of "Large Dam Project" for irrigation and water supply should ensure eight essential activities are well-coordinated.

Lack of coordination in these activities by various executive agencies has adversely affected the benefit cost (B-C) ratio of large dams, thereby giving rise to following suspicions:

(i) Under political and non-technical pressure, the engineers and planners inflate B-C ratio at the design stage, to get the project approved. The design B-C ratio is, therefore, never achieved afterwards;

(ii) Ideal sites for large dams have already been used up. Now only second class and third class sites are being promoted, again under political pressure;

Eight Essential Items for Large Dam Construction

1. Construction of dams.
2. Construction of canals.
3. Construction of field channels and command area development.
4. Watershed development for controlling silt load in the run off water coming into the reservoir and proper land use planning in the watershed area.
5. Rehabilitation of the displaced people from submerged villages.
6. Training of farmers in the command area in economic use of water, in shifting conjunctive use of surface water and groundwater, and development of seasonal lift irrigation schemes for 'out-of-command area'.
7. Infrastructure development like roads, electricity etc. in the command area, for easy marketing of agricultural produce.
8. Establishment of water user groups and promoting women's participation in water management.

In the country, success has not been achieved in controlling salinisation, water logging and reservoir silting and inordinate delays in rehabilitation of dam-affected people. The productivity per cubic metre of stored water has not appreciably increased for the dams constructed in 1990, than their 1950s era counterparts. However, the better line for the country is that large, medium dams and small dams are needed, with coordination of the eight activities and with due care for rehabilitation and protection of the interests of displaced persons.

The tenets of water management policy should include: economical and efficient use of the water in the reservoirs of existing dams, controlling silting of reservoir through watershed development, demand management, recycling and reuse of waste water, prevention of pollution of surface water and groundwater, irrigation of food crops giving more calories per cubic metre of water, use of aquifers having marginal quality of ground water, creating public awareness about the true economic value of water, and of course building bridges of understanding and cooperative action between various stakeholders (Limaye, 2001).

Water Resource Management in the Himalayan region is a great challenge to integrate science, technology and humanity as the mountain system forms the tallest water tower of the world where newer lessons are to be learnt. In this region, the energy exchange mechanisms between the land and the atmosphere are largely altitudinally controlled. In the high altitude region (over 3500 m), the energy releases are in the form of snow avalanches and glacier lake outburst floods (GLOFs) while flash floods and landslides and mudflows assume disastrous dimensions in the altitude range from 500 m to 3500 m. Extreme weather affects smaller watersheds (less than 50 km^2) where flash floods and debris flow are triggered by local heavy rainfall events. Biogeographically, the whole Himalayan region is complex with a sharp and distinct contrast characterizing the eastern warm and humid area with western cold arid region. Roughly, it is estimated that about 20 per cent of the Indian Himalayan region is presently degraded due to natural disasters and man-made problems.

The snowmelt and glacier melt water contribution are estimated to range from 400 to 800 km^3/year. Specific run-off in the mountains is at a maximum in the altitude belt (1500 to 3500 m) of considerable human activity where the estimates of water flow show that on an average, a quantity of 515 km^3/year from the upper mountains (Alford, 1992).

The above water discharges are enormous amounts and all possible efforts should be made to create both small and big water reservoirs for utilization to generate hydropower and other downstream uses. We can follow the cascade development of hydropower adopted by Chinese with the participation of local community. The other significant noteworthy aspects are as follows:

- So far the water resources of arid to semi-arid Indus have been largely utilized. The respective percentages of likely live storage capacities to the average annual flow are 22.2, 10.5 and 0.6 respectively for Indus, Ganga and Brahmaputa river basins demonstrating the need to increase water storage in the Himalayan region. A serious planning is needed to create cascades of reservoirs from the altitude region 3000-4000 m downwards to reduce the turbulent flow of streams and control soil erosion and resulting sedimentation and make water available at all times to all downstream uses for a sustainable development of the region.
- Per capita freshwater resources availability in Himalayan region is evaluated to range from 1,757 m^3/yr in Indus, 1,473 m^3/yr in Ganga and 18,417 m^3/r in Brahmaputra as against an all India average of 2,214 m^3/yr. The excess water from humid Brahmaputra basin could be advantageously utilized by transfer to the semi-arid to arid regions of Ganga and Indus. This would be doubly advantageous as it would reduce floods in Brahmaputra and droughts in Ganga and Indus basins.

- As weather plays an important role in mountains, we should plan several experimental campaigns to better understand the land-atmosphere interactions and develop data bases on snow cover, mountain meteorology and other district bio-geodata for local development under a nationally sponsored programme with regional cooperation and international support for the benefit of human kind and from sustainable considerations for the development of Himalayan region.

10

Epilogue

Himalayas, the highest and the most majestic mountain range of the world, the cradle of Indian civilization, the weather maker of Asia, the source of perennial rivers, which are the life-support systems of the country, have been of unparalleled significance of natures' precious treasure on earth. Magnitude of rock-decay, erosion and destructive mechanisms are presently notably high. We have to learn that periodic adaptability, and change in continuity, are the rules of nature. Majestic mountain chains, its eternal snows and varied vegetation should continue to inspire us for dynamic manifestations and creative activities.

The study of snow and ice is truly an inter-disciplinary science. Basic research in the field is closely integrated with engineering research and technology development in cold regions. For soils, work involves geophysical exploration, geotechnical site investigation, mechanical properties of frozen soils, frost heave, ground water movement and water supply, excavation and drilling, weapon effect, road and airfield pavements and foundation of structures. For snows, work covers snowdrifts, snow removal from roads and runways, visibility problems, road loads, hydrology, vehicle mobility, avalanches and construction on the permanent snow fields. For ice, work involves measurement of ice forces on hydraulic structures in rivers and lakes, construction on and inside glaciers. To utilise the hydro-electric potential in the region, it is necessary to study the Himalayan snow and ice covers in greater details and map the areas with permanent and seasonal snow cover for assessing the volume of water in different parts of the year. The intellectual curiosity of Indian researchers is still to be ignited for the above studies to help manage the cold Himalayan snow and ice regions.

It may be pertinent to point out that the country may need the international assistance for the above. Relevant organisations are International Commission on Snow and Ice (ICSI) which promotes the study of naturally occurring snow and ice and carries out its mandate through scientific investigations into dynamics and physical characteristics of snow and ice systems and the interactions between the Cryosphere, the Atmosphere, the Hydrosphere and the Biosphere and renders expert advise on glaciology to government and non-governmental organisations. It is also responsible for measurements of mass balance at regular intervals. More recently Global Land Ice Measurement from Space (GLIMS) has been launched by NASA, USA in 1999 for providing multi-spectral stereo images of glaciers of high quality. Jawaharlal Nehru University (JNU), New Delhi is already hosting South Asian Centre covering India, Nepal and Bhutan. The efforts to make a meaningful inventory of all snow and ice feeding Indian rivers, needs a priority consideration.

The mountain system is a battleground between conservationalists and commercial interests. Himalayan problems can neither be solved by adopting megabuck, high tech approach nor by doomsayers' dreams that there is no longer any hope at all for the environment. Only a mixture of deep concern and cautious optimism is needed. Avarice and indifference should not be allowed to destroy the natural heritage of tremendous importance for one and all including generations to come for the country. Perennially snow and ice covered regions, i.e. the areas above 3500 m, are largely governed by the natural forces which needs monitoring and understanding while the lower regions are to be saved from excessive exploitation of natural resources and pollution for upkeep of the environmental integrity of the great gift of nature.

Presently, Himalayas are suffering from serious ecological

repercussions, e.g. population increase of diverse cultures; reduction in forest cover, resulting in severe soil erosion, extensive runoff and land degradation. Sub-marginal lands are brought under cultivation resulting in poor soil and water conservation and silting up of drainage channels; shifting cultivation in North-East, dislodging staggering volume of soil; indiscriminate hunting, threatening its fauna and unplanned tourism, creating ecological degradation.

The unique region, being extremely fragile, needs to be managed for its coordinated resource development on a high priority in its own right and not just as an appendage to economic and political interest of Indo-Gangetic plains. The region has a great strategic significance, threatening the integrity of the country from across its borders, particularly China and Pakistan. Backwardness in the mountain region cannot be allowed to continue due to serious repercussions for the whole country like out-migration of population and the fact that the poverty accelerates environmental degradation.

Himalayas should not be treated as a physical entity as it has vibrant living aspects. It can neither be independent of people living therein nor the inhabitants without their habitats. There are areas which are quite rich in natural resources but are inhabited by poor people. The inhabitants have been suffering under harsh environmental conditions, carrying out subsistence economic activities, based primarily on animal rearing and rudimentary farming. This needs upgradation.

We have to remember that in the areas where indigenous people have traditional corporate rights over the resources, the same should continue to be the basis for resource optimization and minimum needs should be satisfied primarily through rescheduling the productive process and not through welfare measures.

A basic task for the scientific community is therefore to observe carefully the natural processes over time, utilizing rigorous techniques for precise measurement to build an empirical basis for understanding. Only such knowledge of trends can we hope to improve future assessments, policies and development programmes in the region.

Let us recall Agenda 21 devoted to managing fragile mountain ecosystems (UNEP, 1992) which calls for integrated watershed development of mountainous areas. It prohibits further degradation, integrated watershead management should conserve and upgrade the natural resource base of land, water, plant, animal and human resources, taking a holistic perspective in the complex processes, including their limits and interconnections. The interested investigators may find some useful information for integrated studies in the Himalayan region elsewhere. (Bahadur, 2003).

Annexures

ANNEXURE-I

GLOSSARY OF TERMS ASSOCIATED WITH SNOW AND ICE

Ablation—All processes by which snow, ice or water in any form are lost from a glacier or snow cover.

Ablation Area—That part of a glacier's surface over which ablation (wastage) exceeds accumulation each year.

Accumulation—All processes by which snow, ice or water in any form is added to a glacier or snow cover.

Albedo—That fraction of incident light reflected in all directions from a surface.

Avalanche—Snow avalanche is the fall by gravity of a mass of snow down a mountain slope. It is effected by the internal cohesion of the snow, the thickness and density of the snow layer, the character of the underlying material and the slope of the terrain.

Basal Slip—Slippage of a glacier over its bed.

Bergschrund—The crevasse which occurs at the head of a cirque or valley glacier which separate the moving glacier ice from the rock wall and the ice apron attached to it.

Calfing—The breaking away of a mass of ice from a floating glacier, ice front or iceberg.

Cirque Glacier—A glacier which occupies a separate rounded recess which it has formed on a mountain side.

COL—Open, U-shape pass across a high, narrow mountain ridge created by glacial erosion.

Compressing Flow—Created by longitudinal compression

in any reach of a glacier experiencing decreasing flow velocity; causes surfaceward movement of ice and an increase in thickness.

Continental Ice Sheet—Sheet of ice of continental proportions, largely burying the underlying landscape and flowing outward in all directions.

Convection—Transfer of heat within fluids by currents resulting from differences in density caused by differential temperature.

Creep—Slow, essentially continuous, non-recoverable deformation (movement) sustained by soils, rocks, and minerals (ice).

Crevasse—A fissure formed in a glacier. Crevasses are often hidden by snow bridges.

Crust—A hard snow surface lying upon a soft layer. It can be formed by sun, rain or wind.

Crystal—Regular, solid, geometrical form bounded by plane surfaces, expressing an internal ordered arrangement of atoms. Aggregates of crystals may have irregular forms.

Dead Ice—Any part of a glacier which has ceased to flow. Dead ice is mostly covered with moraine.

Declination—Difference, at any location, between the bearings to the magnetic pole and the geographic pole of Earth.

Drumlin—Low, streamlined ridge of glacial drift (rock detritus carried and deposited in any mode by glacier) shaped by an overriding glacier.

Equilibrium Line—The line or zone on a glacier where a year's ablation balances a year's accumulation. It is a determined at the end of the ablation season.

Erratic—Rock fragment, typically large, transported and deposited by a glacier.

Esker—Long, low, narrow, commonly sinuous ridge of glaciofluvial material deposited in a subglacial ice tunnel or an ice-walled channel.

Extending Flow—Created by extension within a glacier where flow is accelerating; causes thinning.

Firn—Old snow which has transformed into a dense material.

Firn Line (of equilibrium and snow line)—The line or zone on a glacier that separates bare ice from snow at the end of the ablation season.

Floe—A piece of floating ice other than glacier ice.

Foliation—Crude mineralogical or textural banding formed in rocks, primarily by solid-stage metamorphism.

Holocene Epoch—Past ten thousand years; succeeds the Pleistocene (approximately 1.6 million years of earth history essentially equivalent to the last glacier ice age).

Frazil Ice—Find spicules or plates of ice in suspension in water.

Frost Smoke—Fog-like clouds, due to the contact of cold air with relatively warm water, which appear over newly formed leads, or leeward of the ice edge, and which may persist while new ice is forming.

Glacier—A mass of snow and ice continuously moving from higher to lower ground.

Glacier Flood—A sudden outburst of water released by a glacier.

Glacier Pot-Hole—A deep and more or less vertical hole in a glacier. Glacier pot-holes drain away surface melt water.

Hail—Precipitation of small balls or pieces of ice with diameter ranging frm 5 to 50 mm or more.

Hoarfrost—A deposit of ice having a crystalline appearance, generally assuming the form of scales, needles, feathers or fans; produced in a manner similar to dew (i.e. by condensation of water vapour from the air), but at a temperature below 0°C.

Ice Apron—A thin mass of snow and ice adhering to a mountain side.

Ice Jam—An accumulation of broken river or sea ice caught in a narrow channel.

Ice Sheet—A mass of ice and snow of considerable thickness and large area. Ice sheets may be resting on rock (Island Ice sheet) or floating (Ice shelf).

Icicle—Hanging spike of clear ice formed by the freezing of dripping water.

Insolation—Radiated energy received from the sun.

Kame—Steep-sided a rock or ridge of glaciofluvial debris deposited in contact with glacier ice.

Kettle Hole—Normally bowl-shaped, topographically closed depression within glacial drift formed by melting of a large chunk of partly or completely buried glacier ice, usually filled with water.

Kinematic Wave—Wavelike perturbation of a glacier's steady-state flow that moves outward at a velocity several times normal.

Moraine—Ridges or deposits of rock debris transported by a glacier. Common forms are: ground moraine, formed under a glacier; lateral moraine, along with sides; medial moraine down the centre; and end moraine; deposited at the foot. Moraines are left after a glacier has receded, providing evidence of its former extent.

Moulin—Roughly cylindrical, nearly vertical hole up to 30m in a glacier's surface.

Net Budget—The difference between accumulation and ablation; usually expressed in terms of water equivalent per unit area.

Neve—French term for old granular snow; called firn in German.

New Snow—A recent snow deposit in which the original form of the ice crystals can be recognised.

Nunatak—A rocky crag or small mountain projecting from and surrounded by a glacier or ice sheet.

Ogive—Band or wave on the surface of a valley glacier stretching from side to side and arched in the direction of flow.

Old Snow—Deposited snow whose transformation into firn is so far advanced that the original form of the ice crystals can no longer be recognised.

Outlet Glacier—A valley glacier which drains an inland ice sheet or icecap and flows through a gap in peripheral mountains.

Perennailly Frozen Ground—Continually below 0°C, year after year (Equivalent: Permafrost).

Perfectly Plastic Subsurface—Shows no deformation under increasing shear stress up to a point, after which it deforms continuously without rupture or further increase in stress.

Periglacial—Cold, rigorous environment charaterizing the area peripheral to a large Pleistocene glacier, and the processes and products found there.

Piedmont Glacier—The lobe shaped, expanded, terminal part of a valley glacier spread out over broad lowlands at the base of mountains.

Plucking—Glacial erosive process whereby sizeable blocks of rock are loosened, picked up, and carried away by glaciers (Synonym: Quarrying).

Pressure Melting—Melting of ice where pressure is great enough to lower its melting point below the ambient temperature.

Proglacial—Area lying adjacent to and usually in front of a glacier.

Puddle—An accumulation of melt water on ice surface, mainly due to melting snow.

Rafting—Pressure process by which one floe (a piece of floating ice) overrides the other.

Rime—A deposit of ice composed on grains more or less separated by trapped air, sometimes adorned with crystalline branches, produced by the rapid freezing of super cooled and very small droplets.

River Ice—Floating ice in rivers.

Sleet—Precipitation of snows and rains together, or of snow melting as it falls.

Slush—Snow which is saturated and mixed with water; found on land or ice surface or as a viscous floating mass in water after snowfall.

Snow—Precipitation of ice crystals, most of which are branched (sometimes star shaped). The branched crystals are sometimes mixed with unbranched crystals. At temperatures higher than about –5°C the crystals are generally agglomerated into snowflakes.

Snowline—A line or zone on land that separates areas in which snow remains throughout the year. The altitude of the snowline is controlled by temperature and the amount of snowfall.

Striae (Striations)—Linear, finely cut parallel scratches inscribed on a rock surface by debris carried in basal ice of a moving glacier.

Surge—Relatively short-lived episode of greatly accelerated flow within a glacier.

Valley Glacier—A glacier which flows down a valley.

Varve—Glacio-locustrine sedimentary couplet of contrasting summer and winter laid material, representing one year of deposition.

Whaleback—Smooth, glacially sculptured bedrock knob of modest size resembling the back of a sounding whale.

ANNEXURE-II

HISTORY OF SNOW RESEARCH

Period	*Accomplishments*
Pre-1900	Some observations on snow
Discovery (1900-1936)	Routine measurements on physics of snow, providing insights for study of avalanches and snowmelt stream runoff.
Recent (1936-70)	Government sponsored laboratories established and professional societies founded. Surface energy exchange quantified and physical processes were investigated for engineering applications.
Current (Post 1970)	More sophisticated techniques and tools were applied and significant advances made to resolve field problems but many problems were still to be solved.
Future	Grain-scale geometry of snow quantified for development of models for engineering applications

ANNEXURE-III

GLOBAL ANNUAL SNOWCOVER*

Snow Cover	*Area* 10^6 km^2	*Accumulation* kg/m^2	*Mass* 10^9 $tons/km^3$ *of water*
A. Northern Hemisphere			
(i) Permanent snow on glacier.	2	250	500
(ii) Seasonal snow on land	59	140	8,300
(iii) On pack ice	9	100	900
(iv) On seasonal sea ice	9	120	1,100
B. Southern Hemisphere			
(i) Permanent snow on glaciers	14	160	2,200
(ii) Seasonal snow on land	2	150	300
(iii) On pack ice	5	180	900
(iv) On seasonal sea ice	15	200	3,000

**Source* : Shumsky, P.A., Krenke, A.N. and Zotikov, I.A. (1964): Ice and its Changes: Geophysik of Earth, Vol.II, Washington.

ANNEXURE-IV

GLOBAL GLACIERIZATION*

Glaciarization	***Area*** $10^6 km^2$		***Volume***	
	Accumulation	*Ablation*	*Total*	$10^6 km^3$
A. Northern Hemisphere				
(i) Continental ice sheets	1.1	0.6	1.7	2.7
(ii) Ice sheets on islands	0.2	0.15	0.35	0.2
(iii) Mountan Glaciers	0.1	0.1	0.2	0.03
B. Southern Hemisphere				
(i) Continental ice sheets	13.8	0.1	13.9	28.0
(ii) Mountain Glaciers	0.02	0.01	0.03	0.01

**Sources* :

1. Shumsky, P.A., Krenke, A.N. and Zotikov, I.A. (1964): 'Ice and its Changes', In: *Geophysik of Earth*, Vol.II, Washington.
2. ICEX (1979). *Ice and Climate Experiment*, Greenbelt, Maryland.

ANNEXURE-V

ASTRONOMICAL CYCLES AS PACEMAKERS OF GLACIATION

Astronomically driven changes in the intensity of the seasons might periodically trigger glaciation. This was formalized in 1920s and 1930s by M. Milankovitch, Yugoslav astronomer. Astronomical cycles as pacemakers of glaciation has three components (Broecker and Denton, 1990);

Tilt of the earths spin axis:
Currently, it is at 23.5 degree from the vertical, it fluctuates from 21.5 degree to 24.5 degrees and back every 41,000 years. The greater the tilt is, the more intense seasons in both hemispheres; summers get hotter and winters colder.

The shape of the earths orbit:
Over a period of 10,000 years, the orbit stretches into a more eccentric ellipse and then grows more nearly circular again.

The interplay between the tilt and the eccentricity:
It is the precession of wobble, of the earth's spin axis, which traces out a complete circle on the background of stars about every 23,000 years. Massive reorganizations of the ocean-atmosphere system, are the key events that link cyclic changes in the earth's orbit to the advance and retreat of ice sheets.

*Broecker, W.S. and Denton, G.H. (1990) *What Drives Glacial Cycle'?*, Vigyan Scientific American, Jan 1990 pp. 47-56

ANNEXURE-VI

WORLD DATA CENTRES ON GACIOLOGY

Global Data Centres exist at Colorado, Moscow, Cambridge, Zurich for collection, storage and dissemination of information for the cold regions on the following aspects:

- Studies of seasonal/perennial snow and ice
- Development of materials, techniqhes, observation and understanding of process
- Interaction with the physical environment
- Effect on climate and climate change

There is need to seek regional and international cooperation and collaboration to establish a World Data Centre for documentation adopting established norms and those needed by the countries of the Himalayan region. This has to taken as a priority project for the apex snowy environment due to global warming.

Bibliography

Ahmad, N. and Rais Sarwar (1998) *In : Himalayan Glaciers*, APH Publishing Corporation, New Delhi-110002, p. 150.

Alford, D. (1992) *Hydrological Aspects of the Himalayan Region*, ICIMOD occasional paper No. 18, pp. 68.

Bahadur, J. (1972) *The Himalayan Glaciers*, Science Today July, pp. 21-27.

Bahadur, J. (1981) *Flood Disasters of Glacial Origin*, Proc. Intern. Conference on Flood Disasters, Indian National Science Academy, New Delhi – 3rd to 5th Dec., pp. 15-23.

Bahadur, J. (1983) *A review of Isotopic Techniques for snow and glacier hydrology of mountain watersheds*, Proc. 1st National Symposium on Seasonal Snow Cover 28-30 April, Vol. II, pp. 64-75.

Bahadur, J. (1983) *Some Environmental Problems for Development of Himalayan Water Resources*, Proc. Intern. Seminar on Environmental Impact Assessment of Water Resources Projects, University of Roorkee, Roorkee 12-14 Dec., pp. 82-92.

Bahadur, J. (1993) *The Himalayas: A Third Polar Region*, Snow and Glacier Hydrology (Proceedings of the Kathmandu Symposium, Nov. 16-21, 1992), IAHS Public. No 218, pp. 181-90.

Bahadur, J. (1996) *Eco-hydrological Investigations in the Himalayas*, Proc. Intern. Conf. On Ecohydrology of High Mountain Areas 23-28th March, Kathmandu, Nepal, pp. 13-21.

Bahadur, J. (2002) Hydrological importance of Himalayan snow, ice and glaciers, In: *Where Mortals and Mountain Gods Meet – Society and Culture in Himachal Pradesh*, (ed: Thakur, L.S.) Indian Institute of Advance Study, Rashtrapati Niwas, Shimla, pp. 3-20

Bahadur, J. *Indian Himalayas – an Integrated View*, Vigyan Prasar, New Delhi, p. 279 (ISBN: 81-7480-091-3).

Barry, R.G. (1981) In: *Mountain Weather and Climate*, Methuen London, pp. 313

Bamzai, A.S. and Kinter, J.L.III (1997) *Climatology and and Interannual variability of the northern hemisphere snow cover and depth based on satellite observation*, Rept. No.52 COLA, Md. USA.

Becker, A. and Bugmaun, H. (1996) *Predicting Global Change Impacts on Mountain Hydrology and Ecology*. Integrated Catchment Hydrology/Altitude Gradient Studies, Proc. International Workshop on Global Change Impacts on Mountain Hydrology and Ecology, Kathmandu 30th March-2nd April.

Berri, D.I and Owen L.A (1998) In: 'The role of the Indian Summer Monsoon and the mid-latitude westerlies in the Himalayan Glaciation review and speculative discussion' *Journal of the Geological Society*, London Vol. 155, pp. 353-63 Printed in Great Britain.

Bhatt, C.P. (1992) *The future of large projects in the Himalayas – overcoming Incomplete knowledge and unsound beliefs.*'PAHAR, Nainital, pp.50

Bode, R. (1990) *Asia's Ice-Age Legacy*, Readers Digest 13, No 804, Dec. pp. 41-45.

Chalise, S.R.,(1993) *Regional Cooperation in Hydrological Research and Training in the Hindukush Himalayas*, Snow and Glacier Hydrology (Proceedings of Kathmandu Symposium, 16-21 November 1992) IAHS Publn. No 218, pp. 37-47.

Collins, D.N. and Hasnain, S.I., 'Runoff and Sediment transport from the glacierised basins at the Himalayan Scale' In: *Effects of Scale on Interpretation and Management of Sediment Quality*, International Association of Hydrological Sciences, Publn. No. 226, 17-25.

Department of Science and Technology, New Delhi, *Report of the Study Group on Artificial Augmentation of Snow/Glacier Melt* (Chairman Prof. P.R. Pishorty – July, 1976)

Dhanju, M.S. (1981) *Studies of Himalayan Snow cover from Satellites*, IAHS Publn. No. 145 pp. 401-409.

Dhir R.D. and Singh Harbhajan (ed). (1956) In : *Snow Surveys in the Himalayas*, Central Water and Power Commission, pp. 190..

Evans, S.G., Clague, J.J. (1992) Glacier-Related Hazards and Climatic Change, p. 48-59 American Institute of Physics (AIP) Conference Proceedings 277, *The World at Risk: Natural Hazards and Climate Change*, Cambridge MA, pp. 352.

Ferguson, R. (1984) Sediment load of the Hunza River, *The International Karakoram Project*, Vol. 2, pp. 581-98 ed. Miller, K., England: Cambridge University Press.

Fujii, Y 'Glaciers (in Japanese)' (1977) In Kawakita. J (Ed.), *The Himalaya*, A Sahi Shinbunsya, Tokyo, 136-144.

Geological Survey of India (GSI) (1999) *Inventory of the Himalayan Glaciers – a contribution to the International Hydrological Programme* ed. M.K. Kaul, GSI Special Publication No.34.

Gerrard, J. (1990) Mountain Geoecology – in *Mountain Environments – an examination of the physical geography of mountains*, Publ. Bethaven Press, London, pp. 36-66.

Gerrard, J. (1990) Weathering and Mass Movements, pp. 67-92. In : *Mountain Environments – an examination of the physical geography of mountains*, Publ. Belhaven Press London, pp. 316.

Gillespie, A and Molnar, P. (1995) A synchronous maximum advances of mountain and continental glaciers, *Reviews of Geophysics 33*, 311-364.

The Japan China-Joint Expedition to Qing Zang Plateau, 1989. 'Glaciological climate and Environment on Qing-Zang Plateau', (1993).

Hewitt, K., (1988) European Science in High Asia : Geomorphology in the Karakoram Himalaya to 1939, In: *History of Geomorphology : Autton to Hack*, Tinkler, K.G. (ed.) Binghamton Geomorphlogy Series No 19, Hyman University, pp. 165-203

Hewitt, K. and Young, G.J. (1993) *The Snow and Ice Hydrology Project'*; a Pakistan-Canada Research and Training Programme. Snow and Glacier Hydrology, Proc. of the Kathmandu Symposium, 16-21 November 1992 IAHS Publn. No.218.

International Association of Hydrological Sciences (IAHS) (1994) Publn. No.223, (Proceedings of Yokohama Symposium 12-15th July 1993 on 'Snow and Ice Covers: Interactions with the Atmosphere and Ecosystems'.

Issar, D.P. (2000) *Inventory of Major Glaciers in Indian Himalayas*, A special publication of Survey of India, Geodetic and Research J. branch, Dehradun,.

Italian National Res. Council (CNR) 'The Ev-K2-CNR Scientific Project,' 1990. C/o Department of Earth Sciences, I-20133 Milano via Mangiagalli 34, Italy.

Jacob, K.M. Pennington, W.D., Armbursts, J., Sector, L. and Farhatulla, S. Bull (1979) Scismological Soc. America, 1175-1192.

Jijun Lin, Shuying Xu (1984) 'The contribution of glaciers on the Qinghai-Dizang Plateau and its relationship to atmospheric circulations' In : *International Karakoram Project* (ed. By K.J. Miller) Vol. I, 84-93, Cambridge University Press.

Kanwar Sain (1946) *The Role of Glaciers and Snow Hydrology of Punjab Rivers,* Central Board of Irrigation and Power, Aug 1945 Publn. No. 36 pp.144.

Kelly, P.M. and Hulme, M. (1993) *Climate Scenarios for the Indian Subcontinent,* Report for Climate Research Unit, University of East Augalia, Norwich, Published by International Institute for Environment and Development (IIED), London.

Kotlyakov, V.M. and Lebedeva, I.M. (1998) *Melting and Evaporation of Glacier Systems in Hindukush-Himalayan Region and their Possible Changes as a Result of Global Warming,* Proc. International Conf. on Ecohydrology of High Mountan Areas, Kathmandu, Nepal, 24-28 March 1996, ICIMOD, pp. 367-375

Kulkarni, A.V. (1989) 'Application of remote sensing for glacier inventory and mass balance' Proc. of national symposium on engineering Applications of Remote Sensing and Recent Advances, Indore, pp. 26-30.

Limaye, S.D. (2001) *To Dam or Not to Dam?* – A Water Resources Overview from Developing Countries, Stockholm Water Forum – A forum for global water issues. No.2, June, p.6-7.

Lin, P.N., Thompson, L.G., Davis, M.E. and Mosley Thompson (1995). '1000 years of climatic change in China: ice core $\delta^{18}O$ evidence.' *Annals of Glaciology 21*, pp. 189-95.

Mayewski, P.A. and Jeschke, P.A. (1979) 'Himalayan and Trans-Himalayan Glacier Fluctuations since AD 1812'. *Arctic, And Alpine Research 11*, No. 3, pp. 267-87.

Meier, M.F. and Roots, E.F. (1982) 'Glaciers as a Water Resource', *Nature and Resources,* Vol. XVIII No 1 July-Sept , pp. 7-14.

Meon, G and Schwarz, W. (1993) 'Estimation of Glacier Lake Outburst Flood and its Impact on Hydroproject in Nepal.' Proc. Intern. Symp. On Snow and Glaciers Hydrology, Kathmandu, Nepal 16-21 Nov.1992, IAHS Publn. No 218, 331-339.

Miller, K.J. (ed) (1994) *International Karakoram Project (IKP),* Vol. I and II, Cambridge University Press.

Ministry of Water Resources (1998) Govt. of India *Water Resources Development in India (1947-1997).*

Misra, R.P. (ed) (1992) In: Environmental Ethics – A Dialogue of Cultures, Proc. Symposium, Delhi, March 1-3, 1991, Sustainable Development Foundation, Gandhi Bhavan, Delhi University, pp.168.

Mool, P.K., Wangdev, D., Bajrachaarya, S.R., Konzaang, K., Guyong, D.R. and Joshi, S.P. (2001) *Inventory of Glaciers, Glacial Lakes and Glacial Lake Outburst Floods – Moniitoring and Early Warning Systems in the Hindu Kush Himalayan Region*, Bhutan, ICIMOD, August, ISBN 929115345 pp. 227.

Mool, P.K., Bajracharya, S.R. and Joshi, S.P. (2001) *Inventory of Glaciers, Glacial Lakes and Glacial Lake Outbrust Floods – Monitoring and Early Warning systems in the Hindu Kush Himalayan Region*, Nepal, ICIMOD, August, ISBN 929115331 pp. 364.

Negi, S.S. (1998) *Discovering the Himalayas*, Vol. I and II, Indus Publishing Co, New Delhi.

Niang Yangzhen (1990) *Glacier Water Resources in China*, Janzhoo Glacier Research Institute, Chinese Academy of Sciences, Gansu Science and Technology Publishing House (In Chinese).

Nijampurkar, V.N. and Rao, D.K. (1993) *Ice Dynamics and Climatic Studies on Himalayan glaciers based on Stable and Radioactive Isotopes*, Proc. Intern. Symp. On Snow and Glacier Hydrology, Kathmandu Nov. 16-21, 1992, IAHS Publn. No.218.

Owen, L.A., Wishart, M., Bailey, R.M., Coron, P. and Rhodes, E. (1997) 'Style and timing of glaciation in the Lahul Himalaya, Northern India': A framework for reconstructing lake Quaternary palaeoclimatic change in the Western Himalayas', *Journal of Quaternary Science, 12*, 83-110.

Paterson, D. (1993) 'Did Tibet Cool the World?', *New Scientist*, 3rd July, pp. 29-33.

Paterson, W.S.B. (1969) *The Physics of Glaciers* ,Pergamon Press, Oxford, pp. 250.

Peiji Li (1994) 'Dynamic characteristics of snow cover in western China' In *Snow and Ice Covers : Interactions with Atmosphere and Ecosystems*, Proc. of Yokohama Symp. 12-15 July 1993, IAHS Publn. No 223.

Pé wé, L.T. (1981) 'Permafrost Research in China' Glaciogical Data Report GD-10 'Glaciology in China', World Data Centre A for Glaciology (Snow and Ice) Cooperative Institute for Research in Environmental Sciences University of Colorado, Boulder, May, pp. 68.

Pounder, E.R. (1965) *Physics of Ice*, Pergamon Press, pp. 151.

Purohit, A.N. (1999)'Climate, mountains and environment based development in the Himalayan region' pp. 185-92 In: *The Himalayan Environment* ed. Dash, S.K. and Bahadur J. New Age International (P) Ltd. Publishers, pp. 347.

Samra, J.S. and Raizada, A. (2001) *Biodiversity and Ecosystem Management in Hill and Mountain Agrosystems*. Proc. 5th Agricultural Science Congress, Guwahati, April 4th-7th.

Sharma, M.C. and Owen, L.A. (1996) 'Quaternary glacial history of NW Garhwal, Central Himalayas' *Quaternary Science Reviews* 15, pp. 335-65.

Sharp, R.P. (1988) *'Living Ice – understanding glaciers and glaciation'* Cambridge University Press, Cambridge, pp. 225.

Srivastava, H.N. (1998) 'Earthquakes in Himalayas' Prepublished report, *Vigyan Prasar*, New Delhi.

Stone, P.B. (ed.) (1992) *'The state of the World's Mountains – a global report on behalf of Mountain Agenda*, Zed Books Ltd., London and New Jersey, pp. 391.

United Nations (1992) Agenda 21: *Program of Action for Sustainable Development 1992*. New York: United Nations Conference on Environment and Development (UNCED). UN Publications, Geneva: WMO.

Upadhyay, D.S. (1995) *Cold climate Hydrometeorology,* New Age International Publishers, New Delhi, pp. 345.

Valdiya, K.S. (1998) *Dynamic Himalaya*, ISBN 817,371094 Educational Monographs University Press, Hyderabad, pp. 178.

Vohra, C.P. (1978) *Glacier resources of the Himalayas and their Importance to Environmental Studies*, Proc. National Seminar on Resources Development and Environment in the Himalayan Region, New Delhi April 10-13, pp. 441-59.

Vohra, C.P. (1981) 'Himalayan Glaciers' Chapter 9 of the book *The Himalaya – Aspects of Change*, Ed. By J.S. Lall, India International Centre, New Delhi.

Wake, C.P., Mayewski, P.A., Ping, W. Jiankang, H. and Zichu, X. (1992) 'Anthropogenic Sulphate and Asian Dust signals in snow from Tien Shan, north-west, China', *Annals of Glaciology*, 16, 1992 pp. 45-52.

Wake, C.P. and Mayewski, P.A. (1994) *Modern eolian dust deposition in Central Asia*, Tellus 46B (1994) 3 pp. 220-33.

Williams, Jr. R.S. (1986) *Glaciers : clues to future climate*? United States Geological Survey, Denver, pp. 21

Yamada, T. (1993) *Glacier Lakes and their Outburst Floods in Nepal Himalayas*. Awecs/jica, March, pp. 37

Yamada, T. and Sharma, C.K. (1993) *Glacier Lakes and Outburst floods in Nepal Himalayas*, Proc. Intern. Symp. On Snow and Glacier

Hydrology, Kathmandu, Nepal 16-21 Nov. 1992, IAHS Publn.no 218, pp. 319-30.

Young, G.J. and Hewitt, H. (1990) *Hydrology Research in the Upper Indus Basin, Karakoram Himalaya, Pakistan.* Hydrology of Mountain Areas (Proc. Strbske Pleso Workshop Czechoslovakia) International Association of Hydrological Sciences, Pub. No 190. 139-152.

Index